Searching Smarter

Searching Smarter

John "Skip" Coleman

Fire Engineering
BOOKS & VIDEOS

Copyright© 2011 by
Fire Engineering Books & Videos
110 S. Hartford Ave., Suite 200
Tulsa, Oklahoma 74120 USA

800.752.9764
+1.918.831.9421
info@fireengineeringbooks.com
www.FireEngineeringBooks.com

Marketing Coordinator: Jane Green
National Account Executive: Cindy J. Huse

Director: Mary McGee
Managing Editor: Marla Patterson
Production Manager: Sheila Brock
Production Editor: Tony Quinn
Cover Designer: Karla Pfeifer
Cover Photo by Lt. Bob Pressler, FDNY (ret.)

Library of Congress Cataloging-in-Publication Data
Coleman, John, 1950-
  Searching smarter / John "Skip" Coleman.
      p. cm.
  Includes index.
  ISBN 978-1-59370-258-8
  1. Lifesaving at fires. 2. Search and rescue operations. I. Title.
  TH9402.C65 2011
  363.37'81--dc22

                    2011007442

This book is dedicated to Diane Feldman, managing editor of Fire Engineering *magazine. Although she has never worn "the uniform," she works tirelessly to help educate the nation's fire service and to constantly present us in a positive light. She has become a great friend to me and my family as well as to the fire service family.*

It is also dedicated to my beautiful wife Theresa. She has put up with me and the same old jokes for more than 31 years, and I can only pray she continues to do so for another 30 or so. To my son Toby, his wife Victoria, and my daughter Fay.

Lastly, this book is being completed when the nation and the world are changing at an alarming pace. God help us and be with us all! The republic has worked well for more than 200 years; I pray it continues on as the founders envisioned.

Contents

The Scope of the Problem 1

I remember starting my career as a firefighter back in 1975 and all the training that I experienced in recruit school. When I think about it now after 32 years, the subject of search was about as important to the curriculum of my recruit class as it was for the curriculum of my entire fire service career.

The thing that sticks out the most in my mind about recruit training was laying and pulling hose. It seemed that for hours on end we would connect to a hydrant, lay out 500 ft. of supply (3-in. and 2½-in.) hose, connect the hose to the engine intake, then pull a preconnected attack line. For days and days we would pull and stretch hoseline. We also spent many, many hours raising 35-ft. ground ladders. We did right-angle raises, parallel raises, three-firefighter raises, four-firefighter raises. We did raises we knew we would never use again after drill school—the dome raise comes to mind.

We spent time on knots and on self contained breathing apparatus (SCBA). We tied them and untied them, donned them and doffed them. We did all this for hours and hours. With all those hours we spent, it's sad to say that my best recollection of search training consisted of part of a single afternoon.

That was way back in 1975. In 1984 I was tasked with giving detailed training as an instructor for a class of 64 recruits. My assignment was to train the recruits in SCBA and search. I was given a 40-hour block of time to train each group of eight recruits in the use of SCBA and how to conduct a primary search. Since 1984, to the best of my knowledge, every recruit in Toledo has been given at least 40 hours of SCBA and search training. After recruit school I can think of no formal, organized, department-wide search training that was ever conducted before or after 1984 with one exception. While teaching the recruits of the 1984 class the oriented search, deputy chief Robert Schwanzl sat in one of my "hands-on" search sessions. He was so impressed with the evolution that I was ordered to stay in training after the recruit class graduated and put all Toledo firefighters through the oriented search training.

Where We Are Today

When I think of every evolution that we do on the fireground, search is the most poorly performed. Most fire departments are great at pulling and stretching hoseline. We can stretch line forward, backward, upstairs, downstairs, through windows, and out windows to buildings next door. We can take hydrants with forward lays and reverse lays. We can tandem pump, draft, inline pump, and pump from the tank.

Today when we can't think of any more pump drills for the morning drill session, we go out and throw ladders up against the station: three-firefighter raises, two-firefighter raises, four-firefighter raises, both parallel and right angle. We do all that, but I guarantee you that most officers and firefighters never practice searching. Many leave the firehouse without even a basic plan of how they are going to search the building in front of them at their next fire.

As I said, we perform search the worst of all our evolutions on the fireground. I don't think most of us are searching the way we would want someone to search for us or for our family members. When we search, we often miss the center of a room. We miss corners. Sometimes we miss rooms altogether. Other times we search where only dead people could be. In most instances, especially in larger buildings, we take way too long to search.

One of my concerns in writing this book is a general lack of defining who will be tasked with conducting a search. Many proposals for classes and seminars cross my desk. Some of these are directed toward search evolutions. Many of those proposals directed for search include the words "truck company." A high number of others focus on "large area search." I have concerns with both of these groups of seminars and training classes.

First off, not everyone has a truck company. In the past and with larger fire departments, it was standard operating procedure that truck companies conducted searches. There are 200 metro departments recognized by the International Association of Fire Chiefs. That's 200 out of 32,000 departments in the United States. The vast majority of fire departments in the United States today respond to a typical house fire with 13 or fewer firefighters. Many of these departments don't respond to a house fire initially with a truck. Of those departments that do respond initially with a truck company, many may respond with two or fewer firefighters on board.

Second, the concern I have with large area search centers around the aspect of time. This will be discussed in greater detail later in the book, but let me emphasize here that preparing for and conducting a search in a large area, for example a large chain restaurant, may take a single crew from 45 minutes to an hour to conduct. How many viable victims do you believe you will pull out of that restaurant after 45 minutes in an immediately dangerous to life and health (IDLH) atmosphere?

Common Search Mistakes

Let me start this section by stating that no one individual or department is perfect. Additionally, there is almost always more than one right way to fight a fire or to conduct a search. But after reading countless accounts of fires in trade magazines along with OSHA and NIOSH reports and on the Web, I believe many mistakes are made while conducting searches. These are happening not only in large commercial occupancies, but in single-family residential homes as well.

Lack of prior planning

One of my biggest concerns in the way searches are conducted is that as today's fire officer walks toward a fire building preparing to conduct a search, the furthest thing from the officer's mind is developing a plan. I have asked literally thousands of firefighters during training sessions and seminars if they first developed a search plan before they approached a building on fire with the intent of conducting a search. Unfortunately, I didn't get too many answers in the affirmative.

I don't understand this. As firefighters, we are trained to think on our feet and develop action plans literally in seconds as we prepare for a specific evolution. If assigned to fire attack on the second floor of a two-story wood frame house, most officers will have developed a plan including the size of hose to be pulled, where the hose will be stretched, and the type of knockdown application to be used, all before getting out of the jump seat. The same holds true for an officer tasked with ventilation. Again, glancing at the fire building, the officer quickly evaluates the type of ventilation—horizontal or vertical—and a means of opening up the building, if required. Additionally, the location of any opening is evaluated prior to the pulling and positioning of ground ladders.

Again I ask you the last time you were assigned search, what type of plan did you devise prior to entering the building with your crew? Hopefully, by the time you finish this text, you will learn to develop a plan for every search you are required to conduct.

Lack of focus!

If you've heard any of my lectures or read any of my previous books, you'll know that a word that I often refer to is a "focus." I believe the incident management system is built on focus. It is the incident commander's job to focus on the entire incident. It is the job of group and division officers to focus on their specific assignment.

As a search officer, you have several areas of focus that must be maintained in order to conduct an efficient search while maintaining the safety of your crew. An officer assigned to search must maintain focus on the following.

Building construction. It seems that more and more firefighters are been involved in floor collapses. Houses built prior to the mid-1980s were, for the most part, constructed with floor assemblies made with 2×10-in. floor joists or greater. These assemblies were certainly more resistant to the effects of fire than the newer lightweight floor assemblies constructed since the mid-1980s. The search officer, as part of initially sizing up the building for search, must consider the age and specific construction features of the building to be searched. Additionally, as the search is being conducted, the officer must continually evaluate fire conditions to ensure that this structure is this safe as possible.

Fire conditions. The search officer must constantly maintain an awareness of the fire conditions, not only in the area being searched, but also throughout the majority of the structure. This may require a radio transmission to the attack officer or commander to get an update on fire conditions. Civilians can only tolerate so much heat, heavy smoke, and fire conditions. Your search plan should take these aspects into account when prioritizing search areas.

Your search plan should include where you want to start and stop the search. If you believe the search area has become threatened by fire, you may need to re-evaluate the route.

Progress of the search. Assuming zero visibility (as assumed throughout this book), unless searching a one-room structure, the search officer needs to be constantly aware of the search crew's progress. This is especially true if your search crew has more than one member, which I certainly hope it has. The search team has a lot of area to cover in a small amount of time. In order to most effectively utilize searchers, the search officer needs to know which searcher's area has been completed first. It is then up to the search officer to lead that firefighter to the next search area.

Lack of search continuity

Search continuity is defined as *assuring that the search is conducted in a logical and uninterrupted manner*. It entails knowing where we have searched, where we are searching now, where we will search next, and when the entire area has been covered, even if we find and remove the victims.

The officer conducting a search where no victims are located generally has a pretty easy assignment. Most of the searches we conduct come up negative. Maintaining search continuity in these instances is a relatively simple task. Pick the appropriate place to start, move logically throughout the building, and end up at your "stopping" point. On those rare occasions when we do find victims, it is generally where search continuity is disrupted. There are four things that must be known and kept in mind when a victim is located and is being removed, in order to maintain search continuity:

- Where the searcher(s) have already searched

- Where the victim(s) were found

- Who is taking the victim(s) out

- How to return to where you left off to continue with the search.

It is imperative that the search officer is constantly aware of these four items in order to maintain continuity of that specific search. This awareness will help to ensure that all areas have been searched at least once and no time has been wasted in searching areas multiple times. Maintaining search continuity will also help ensure you don't miss areas entirely.

What This Text Will Cover

This text is divided into four sections. Section 1 is an introduction to search. There I discuss the different types of search used in today's fire service. Additionally, I discuss the incident commander and search officer positions and how they can act together to most effectively locate and remove endangered victims. Finally, I discuss the search officer's relationship with other group and division officers.

Section 2 discusses search in detail. To the best of my ability, I get into the specifics of the four known types of searches being used today.

Section 3 covers search basics. This will undoubtedly be the most controversial section of the text. This is "search according to Skip." Not everyone will agree with everything I say in this section. Take what you can work with from this section and leave the rest.

The last section discusses search in specific occupancy types. Both residential and commercial occupancies are discussed in general, as well as how those environments specifically affect a search.

Types of Search Used Today 2

I'm not sure how my dad was taught to search when he came on the job in 1952. If my suspicions are correct he probably wasn't formally taught search at all. More than likely, on his first job where he was required to do a search, his officer said, "Hold on to me and don't let go!"

Hopefully they talked about it when they got back to the station. On the next job where he was required to do a search, he was given direction by his officer and again, afterwards, they discussed what they did and why they did it. Eventually, after doing this several times, a pattern began to develop in his mind toward what a proper search entailed.

As I mentioned in chapter 1, I was sent to the training bureau in 1984 to train a recruit class of 64 young firefighters. My assignment was search and SCBA utilization. Unbeknownst to me, this was the beginning of a long journey for me in the subject of search. My first problem was that I hadn't had a lot of formal training in the proper search evolutions. I don't even know if I was aware in June of 1984 of any specific name for a type of search. I started reading trade journals and textbooks in order to develop a lesson plan that would be suitable for the recruits and, even more importantly, would show them a solid, safe, and efficient search method that they could use when they hit the line. My final concern was in developing and teaching a search evolution that could be understood by incumbent firefighters when these 64 recruits were finished with recruit school.

After many "over-the-desktop" conversations with the other 11 instructors and the information gleaned from the trade journals and textbooks, I developed the lesson plan that is still being used today by the Toledo Fire Department. Since 1984, I have traveled across the United States teaching hundreds of fire departments in a wide array of topics. Most of those lesson plans contain at least a small section on search.

I always ask anyone in attendance if they are familiar with any other types of search. In the 25 years that I have been teaching, I have never had anyone give me a fourth method. As you will see in section two, I list four types of search. The fourth method that I discuss is vent/enter/search (VES). Although I do not consider the original VES a specific form of search but rather a variation of two forms, I felt due to its misunderstandings and dangerous components that it called for examination in detail.

The four types of search

I'd like to think that if I'm anything, I'm a realist. I realize that there is almost always more than one right way to fight a fire. Firefighters as a whole are often too eager to criticize others' actions at fires. It is my belief, backed up by my responses to thousands of fires in my 32 years of service, that there are too many variables present at too many locations at a single fire to think that only one course of action will be effective. What looks like a 1¾-in. hoseline fire attack from the front may also present all the qualifications for a deck gun application from side C. These are both viable attack options at a single fire.

The opposite also holds true. There is almost always more than one wrong way to fight a fire. It is hard to defend the pulling of a booster line into a well-involved, 10,000-sq.-ft., vacant auto repair facility.

It is said that variety is the spice of life, and with variety comes options. Let's look at some of the options available today for a simple room-and-contents fire. We can work off of booster tank or from a hydrant supply. We could pull a 1½-in., 1¾-in., or 2-in. handline. We could use a direct or indirect form of water application. We could vent with natural or mechanical means using an axe or fans. The proper choice in these options is dependent upon the location and size of the fire and a multitude of other variables, including on-scene staffing and apparatus types.

I am aware of four types of search. Not one of these is a panacea, but I do believe that one of them is by far the safest and most effective form of search. Each has its positives and drawbacks. The key in my mind is to become familiar—no, competent—with all forms of search and their variations in order to best protect the lives of civilians and brother and sister firefighters when they get in trouble. I look at each as a tool in the toolbox. A good carpenter or mechanic should be able to use all of tools in a safe and effective manner. Sure, there may be a "go to" tool to use whenever possible, but when only one tool can do an effective job, the worker can grab it with confidence.

- The four types of search that I will expound upon in this text are:

  ▷ **The standard search.** This is probably the form of search you learned in recruit school. The standard search consists of a crew of at least two members who enter the building and search one room at a time.

  ▷ **Team search.** Team search is utilized in large area buildings. My definition of large area buildings are those buildings with single areas of more than 3,000 sq. ft. This definition is certainly up for debate and not all-inclusive. Houses and small, older commercial buildings that have been converted to restaurants certainly fall into the criteria that could require team search, but fail to meet my initial definition. Team search utilizes a rope as a search tool that provides the searchers the ability to maintain an awareness of their egress point.

   ▷ **Vent/enter/search (VES).** The original vent/enter/search is defined as an emergency means of searching second-floor bedrooms off of a porch roof for probable victims, using a roofing ladder for access to the second floor from the exterior. The original method is extremely dangerous. However, when two firefighters use this method combined with the oriented method of search, it becomes a very effective and safe form of search.

   ▷ **The oriented search.** The oriented search is a method of search that divides responsibilities into two specific areas. The officer's responsibility is the safety of the crew and developing and following the search plan. The searcher's responsibility is to conduct the search.

Urban search and rescue

When I discuss the three types of search used today, I am speaking in the context of structural firefighting. As I said in chapter 1, this text will look at search and how it can be applied to the different occupancy types that exist in the United States today. I realize that there are other forms of search that are applicable to firefighters that go beyond the scope of structural firefighting. Among these are urban search and rescue (USAR), which is used in collapsed and structurally compromised buildings. Urban search and rescue is a very specialized form of locating victims in buildings that have collapsed, either partially or completely. Urban search and rescue is as much about shoring and cribbing and other forms of stabilization as it is about actually searching for yet-to-be-located victims.

Wilderness and urban searches

Firefighters have been called to assist park rangers and law enforcement agencies in searching for lost civilians in wilderness and urban areas. These forms of search are generally conducted during daylight hours and under less hazardous conditions. Weather may be a contributing factor during some of these searches, especially during extremely cold weather, but wilderness searches are usually less hazardous to the searcher due to better visibility conditions (i.e., the absence of smoke and presence of daylight). Searchers are generally given boundaries in which to search during wilderness and urban searches.

The Incident Commander and the Search Group

3

I have always believed that the incident command system is built on focus. The incident commander has specific responsibilities of focus. Subsequently, the assigned group and division officers also have their specific responsibilities or areas of focus. If everybody from the incident commander down to staged companies focuses on their specific responsibilities, then everything gets done.

The Responsibilities of Command As It Relates to Search

It is the responsibility of the incident commander to prioritize when search operations are to begin. There is no easy rule of thumb to follow here. Where we place search on the list of priorities depends on several factors. Experience probably has the greatest influence on an incident commander's ability to correctly prioritize when search should begin, or if search should begin at all. An actual fire in an occupied residence is not the time or place to practice trial and error. Only with dozens, if not hundreds of fires under his belt will an individual gain the experience necessary to correctly decide when and if a primary search will begin.

- There are several factors that need to be weighed when determining the priority of search at any specific fire. These include:

  - **The probability of savable occupants being inside.** Later in this text, I will get into specifics about prioritizing a building for search, but an experienced incident commander should be able to look at a building involved in fire and make some educated assumptions concerning the likelihood that savable victims still could be inside the structure. I emphasize "savable victims." This relates to the first statement in the National Fire Academy's risk policy. We will take great risk to save life! Other than firefighter safety, where an incident commander sizes up a fire initially to determine if firefighters can safely and effectively enter the building involved in fire or not, civilian safety or the likelihood that savable victims could still be inside the structure plays the second biggest role in determining when or if a search will be conducted. If there is a chance that savable victims still can be inside the structure *and if* firefighters can safely and effectively enter the structure and become part of the solution rather than part of the problem, then search efforts should be relatively high on the prioritization list. The converse is also true. If there is no chance that savable victims could still be inside the building involved, or if firefighters themselves wearing their full protective gear cannot safely and effectively enter the structure because of fire conditions, then the search, if conducted at all, would be a secondary search after extinguishment efforts have been conducted.

▷ **The size of the fire.** A small fire provides the greatest chance of survivability for civilians remaining in the structure. Again, the converse is true. The larger the fire or the more of a building that is totally involved, the less the chance of survivability for any remaining occupants.

▷ **The location of fire.** The general rule of thumb is the lower the fire, the greater the problem. A basement fire presents a greater number of potential issues to arriving fire crews than an attic fire. The avenues and potential for spread of fire and smoke are significantly greater in a basement fire than a fire in the attic.

▷ **The location of fire as it relates to the location of occupants.** Not only is it important to determine the location of the fire as soon as possible, but the incident commander should try to rapidly determine the location of the fire as it relates to the location of the occupants. Some assumptions may have to be made here. The time of the fire is one factor. Sleeping areas may be given a higher priority in search at late-night or early-morning fires. Conversely, living and cooking areas may be given a higher priority during daytime and early evening hours. Because of these assumptions, search may not be prioritized quite as highly in a fire involving the second floor of a two-story house during the daytime or early evening hours. This is not to say that children never nap in their second floor bedrooms in the afternoon. As I stated above, there are no hard and fast rules but only generalities that can be considered rapidly during the prioritization process.

▷ **Occupancy type.** This text will deal with many occupancy types as it relates to search. Again, no hard and fast rules should be made, but educated generalities can be drawn upon. As an example, my priority for searching a strip mall with a fire in an ice cream shop at 3:00 a.m. would be fairly low. My priority for searching a restaurant with a full parking lot at 8:00 p.m. on a cold Saturday evening would be extremely high. A residential occupancy should be a high priority for search as long as the building appears to be occupied (that is, people are living in the structure). Know the buildings in your district or community, including the non-residential buildings. Get out of the chair in the afternoon and drive your district and note which apartment buildings or complexes are occupied or vacant. I am not saying that every specific apartment be inspected for occupancy, but every officer and firefighter assigned to a company should know which apartment buildings are totally vacant. The same goes for the commercial occupancies in your district.

▷ **Staffing on-scene.** The last factor that I will discuss at this time concerning prioritization of search is the amount of staffing that is on-scene when consideration is being given to assigning search. There is some math involved here along with the ability of an incident commander to realize exactly how many firefighters it takes to do each task. These numbers are specific to every department and depend upon the number of members assigned to an apparatus, their training, and the number of apparatus dispatched and on-scene at the time search is given consideration for assignment. How many firefighters will it take to get a line on the fire? How many firefighters will it take to commence ventilation efforts? I suppose the greater question is how many firefighters will it take to search a large restaurant in a reasonable amount of time with zero visibility?

I will offer a rule of thumb that will permeate this text: If staffing is low and you are torn between putting out the fire and searching and only have enough staffing to do one thing, put the fire out! One or two on-scene crews can only do so much. Another way of saying this is that six or fewer firefighters can only do so much in a given amount of time. If you pull up to a fire situation where chances are that people are inside and in jeopardy, but you only have the time and staffing to do one thing and do it with the most positive effect on the outcome, then pull a hose line, head for the fire, and put the fire out! Doing this will eliminate or diminish the majority of your initial problems. If you stumble upon a victim while accomplishing this, you have one terrible decision to make, especially if it is believed that more than one victim is in that area of the fire building.

The Relationship Between Command and the Search Officer

The Incident Command System (ICS) is a wonderful tool, and I believe it has made the life of the chief officer or incident commander much easier. As with any tool, the user has to trust the tool and feel confident in his or her ability to use it. It was not uncommon 30 years ago, prior to ICS, that a fire chief (incident commander) would hold his breath and quickly run into a burning building to check on the progress of his interior crews. I hope today's incident commanders are smarter than that. Couple those brains with technological advances, more specifically portable radios, and the person running a fire can remain outside to focus on the entire incident. There has to be a trust or a bond between the incident commander and interior group and division officers that allows them to operate in the safest and most efficient means possible. This requires that the incident commander make some assumptions concerning the search and the officer running the search:

- The search officer actually has developed a plan for conducting the search. Part of this is what I call "reading the building for search." I will discuss this more in later chapters, but it is unconscionable to think that a company officer assigned to search would not have developed a plan for where to begin, move through, and end the search when confronted with a fire situation.

- The search officer is, as much as possible within the constraints of staffing, coordinating the search effort. My utopia is that the officer assigned search rarely if ever sweeps under beds and in corners conducting an actual search. That is not the officer's job. Using the right search at the right time allows the officer to coordinate a logical search path in the least amount of time, while assuring that all building areas immediately dangerous to life and health (IDLH) have been searched. I am of the fervent opinion that a search officer cannot actively search while simultaneously coordinating a search plan and do both effectively. Something will be short-changed, and it is in this particular instance where nothing can afford to be cheated.

- The search officer will keep personal safety and crew safety foremost in mind. While not attempting to beat a dead horse, I want again to emphasize the relevance of focus. It is the officer's job to focus on the entire search, knowing where to search first, last, and everything in between. It is the searcher's responsibility to actually sweep under beds and in corners. In doing so, we divide and conquer the responsibilities and everything gets done.

- The searcher must inform command when changing floors. There have been many articles and textbook chapters written about firefighter accountability systems. I do not advocate the use of any system for accountability that needs to be purchased, above and beyond what you already have in portable radios. In my opinion, accountability is a communication process. Prior to entering the building for any assignment, the officer should inform command by radio of the unit's designation, current assignment, and where the crew's operation will begin in the building ("Engine 5 is assigned search, starting on division two"). This simple communication assures that the incident commander and the officer are singing from the same songbook. Once that entire area has been searched, I expect the officer to inform command that the area is all clear and the unit is moving to another area or level of the fire structure ("Search to command: we give an 'all clear' on division two; we're going up to division three").

- If the search officer splits the crew for any reason, the officer will tell command. This goes back to accountability. There are only a few reasons why a search officer would split the crew. If the officer decides to split the crew into separate search groups, command needs to be informed in order to track the location of each crew.

- If search believes there are still savable victims in a building or area that the crew cannot cover in a reasonable time, the officer will inform command of that fact.

The search officer also has to make a few assumptions concerning the incident commander:

- If the incident commander receives information concerning the search, the commander will relay that information to the search officer. This may seem like a no-brainer, but I have heard several accounts where the incident commander gained information concerning the number of victims, their current location, or similar issues and neglected or delayed relaying the information to the search officer. Time is of the essence in search and anything that can expedite or create a more efficient search could mean the difference between life and death.

- Command will assign the backup group as soon as resources are available. As you will later learn, I believe that at any residential occupancy fire the search crew need not have a hoseline with them. Because I also do not like to send a crew into a burning building without the protection of a hoseline, I try to assign a backup crew to shadow the search group as soon as the resources are available. My priority for this assignment depends on several factors, including amount and location of the fire, relative location of the victims to the fire, and if the fire is located on a floor away from where the search is being conducted.

- Command is watching out for changing fire conditions.

One of the most challenging and stressful situations between the incident commander and a search officer is when victims are found. First and foremost, this information must be relayed to command as soon as a victim is located. If the victim is obviously deceased, then a brief radio transmission should be made to advise command, using the fire department's unique terminology for the situation ("Search to command: we found a victim who is coded"). If a possibly viable victim is located, this information should also be transmitted to command with the search officer's intent ("Search to command: we've located a victim and we're bringing the victim out the front door"). Command can now prepare to receive the victim by assigning an EMS unit to be staged at the front door and await the victim. If there are believed to be several more victims inside the structure, command could also grab an assigned rapid intervention team (RIT) unit and instruct them to meet the search crew at a specific location ("Command to search: I'm sending RIT to the bottom of the stairs to grab the victim; continue on with your search when they get the victim").

The incident commander needs to make these decisions based on experience, information gained by looking at the whole picture, and the information received from division and group officers. Command should have the overall "big picture" of the scene, including the staffing available, what has been done, and what still needs to be assigned.

Search and accountability

NPFA Standards 1500 and 1561 both have wording related to accountability. Both sections make the following points:

- It is the incident commander's responsibility to maintain an awareness of the *location* and *function* of every crew on the scene. Those two little words—location and function—should make any existing or potential incident commander shiver. It is not enough simply to know who is in the building and who is not. Location means location *in the building*. In a house, that would mean division one or two. In a big house, it could be division one, quadrant C, as examples. In an apartment, it's division and apartment number. In a strip mall occupancy, it's occupancy name and quadrant. In a warehouse, I honestly have no idea, but that's no excuse. Now be honest with yourself: do you believe your procedures provide a means for the IC to know the *location* and *function* of every crew on the scene of a fire at all times? There's a saying in basketball, "no harm, no foul," but mark my words, with the staffing reductions and litigious atmosphere that permeates today's society, more and more fire chiefs and incident commanders find themselves on the receiving end of lawsuits or even criminal charges. Negligence is an ugly word. The standard is out there and has been for years. Ignorance is no excuse.

- It is the company officer's responsibility for making command constantly aware of the crew's location and function.

- It is the crew member's responsibility to make the officer aware of the member's location and function at all times.

So, what does this mean, how does it relate to search, and how can it be fixed?

It means every department must have policy and procedures in place to ascertain the location and function of every crew on the scene. If you don't, you are leaving yourself wide open for trouble, not to mention the fact that this improve the safety of your crew members.

How it relates to search is that one of the officer's key responsibilities is to communicate to the IC the crew's location as they move throughout the building. To achieve this takes only a portable radio. Communication is the key. As the crew enters the building, I expect the officer to give a brief announcement to command stating the crew's designation, assignment (*function*) and where operations will begin (*location*) ("Squad 1 is starting search on Division 2").

As the crew moves through the building, their location must be updated as required. In residential occupancies, this is generally the division or floor of operation. ("Search to command: All clear Division 2. Moving to Division 1.")

The same should be done for all groups and divisions working on the fireground, including staged crews. This communication should be continued until the fire is deemed "under control."

The Relationship Between Search and Groups/Divisions   4

If you've read any of my other material concerning the incident command system, you'll understand my attitude toward focus. I believe that ICS is built on focus. It's "divide and conquer." The incident commander is responsible for dividing the fireground into manageable units. Where I come from, that is called sectoring. Once a fire crew is given an assignment by the incident commander then it is that officer's responsibility to focus on the specific assignment given.

Attack puts out the fire. They do not vent the building. They should not search the building for victims. If they stumble upon a victim while conducting a fire attack, that's a whole different scenario. That said, I don't believe the most effective fire attack can be mounted if the attack crew is searching for victims as they advance the hoseline. Communication is the key. If the attack crew stumbles upon a victim, that fact must be immediately transmitted to command, who will then make a reasonable decision based upon the information at hand.

Scenario 4–1. Engine 5 is assigned attack. As they advance a 1¾-in. hoseline to division 2 toward the seat of the fire, they find a victim at the top of division two stairs. I would expect the following radio transmission from the attack officer:

"Attack to command, we found a victim on division two. We're going to bring him down."

Command now has a decision to make. By saying "command OK," the commander believes the best option is to suspend a fire attack and allow the attack crew to bring the victim out of the building. However, command may have a crew outside all geared up and ready to meet the crew and bring the victim out while the attack crew maintains their advance on the fire.

Search and Other Groups

Now we will look at the groups assigned most at working fires and discuss the relationship between the search officer and the officer of the other group.

Search and attack

An attack group is generally the first assignment made at the vast majority of fire responses. Experience has taught us that the faster we put out a fire, the fewer problems we have to overcome. In my 32 years on the job, I can only remember a handful of fires where attack was not the first assignment made. Most of those were assigning "Exposure" at fires that were well advanced and spreading to other structures.

The fact that attack is generally assigned before other groups and divisions gives us a big advantage in the protection of the search crew. Under most circumstances the attack crew will pull and stretch all line of appropriate size to darken the fire. The line is usually taken in by the most appropriate route in order to extinguish the fire. This puts a line between the fire and savable victims. It also puts a line between the fire and savable property. If done correctly, both of these actions should produce a positive outcome.

In doing so, two other advantages surface for the search crew. First, it places a line between the fire and the search crew who, in theory, should be behind the attack crew. Second, it provides the appropriate avenue, in most circumstances, for the search crew to enter the building. Attack should enter the building from the best vantage point of cutting off the spread of fire. As you'll soon learn in subsequent chapters, search should start as close to the fire as possible where savable people could be located. The search officer generally should follow the initial hoseline into the building to a point where it's unlikely savable victims would be found, then should commence the search.

I don't believe Attack can extinguish fire in hidden areas *and* search for viable victims at the same time if they want to do either properly. Likewise, I don't like the concept of the search crew searching for both life and fire, as some departments advocate. If the attack crew finds a victim, then I expect them to notify command, who will then determine how that victim will be removed from the building. If the search crew locates fire that has not been attended to by an attack group, then I would expect the search officer to provide that information to the incident commander who again will determine how and by whom that fire will be extinguished.

Command has several options available to handle the situation. The commander can simply inform the attack group that search has located a fire in a specific area and Attack can proceed to that area when they get a chance. Another alternative could be that Command assigns an entirely different crew to handle the fire that Search located. The size of the fire Search finds, along with its location, are deciding factors in how the incident commander will react.

Search and ventilation

As the search group moves throughout the building, I expect them to have almost 100% of their focus on searching. I don't expect the search officer to allow the crew members to open up for ventilation purposes as they go along. The context of this book is not ventilation; however, some basic principles apply here. The most basic is if a crew member or the search officer feels the need to open a window while searching due to heat or fire conditions, I believe they are searching in the wrong area. How many unprotected civilians will they find alive in such conditions? Second, hot gases and gases under pressure always flow from areas of high pressure (the fire) to areas of low pressure (that now-opened window). This fact could allow heat and fire to be drawn toward the area and window that the search crew member has just opened with possible negative outcomes for the search team.

Today's ventilation techniques rely on certain technical factors. Indiscriminately opening up windows and other openings in a fire building may be counterproductive to the planned ventilation efforts. Before any opening other than the interior doors is made by the search crew, I believe a transmission via radio to the vent crew (if assigned) should be made stating the searchers' intentions. This gives the vent group officer the ability to control the ventilation efforts. Here again I must emphasize that if heat and fire conditions dictate that windows must be opened to improve the conditions where the search is being conducted, then the search crew has bigger problems than just looking for victims. If you're not comfortable in your $2000 bunker gear because of heat present in an area, do you really think there are savable victims nearby? The answer is almost assuredly "no!"

Search and backup

The backup group is a luxury for the incident commander. At times this is a dangerous job and staffing is not what it used to be. Most departments are responding with 13 or fewer firefighters to a reported fire. If, upon arrival, an actual fire is present, then the task of the incident commander is to assign on-scene crews to handle the most pressing problems. The available staff should stay focused on that assignment until the problem is eliminated or diminished to the point where another assignment can be given to that crew. That is called prioritizing.

Most incident commanders prioritize attack first, ventilation or search next (depending upon a variety of factors), and then finally backup groups to provide protection for interior crews. Those last assignments may only happen as staffing becomes available, so you can't always count on having a backup group assigned. This may not be the safest way to fight a fire, but it is the way most of us have been taught and accepted as the norm.

The search officer must keep crews informed as to their current location. In a house fire, this is tracked simply by the floor or division where search is operating. This way, when command has the staffing to assign a backup group, they should know that search is on a specific division of the house and without the protection of a hoseline.

This process becomes much more complicated in commercial occupancies. I will not get into the specifics at this point, but you will find that certain occupancy types such as strip malls are not that complicated when determining the location of the search group. This is not true of a warehouse, large mercantile occupancy, or even a large restaurant.

The responsibility of backup for shadowing search is to provide a protective hoseline slightly remote from the area being searched, so as not to slow down or inhibit the search efforts, and to assist in a rapid and safe egress from the fire area if fire conditions require. As with any group or division that backup is assigned to, their focus is not supposed to be locating fire to extinguish. If their focus is in finding fire to extinguish, how much effort are they putting into protecting the crew they have been assigned to back up?

There are almost always exceptions to everything. Notice I said *almost* always. When I envision a search team conducting a search, in my mind the officer and searchers are focused on conducting the search. I don't expect them to open up walls to check for fire or windows to let out smoke. As with other groups and divisions, communication is the key. If you open up a closet door as you search to possibly locate a child in hiding, and find fire in the closet, that fact must be communicated to command.

If you have finished searching the structure—let's say it's an apartment building— and you are on the sixth floor and encounter only light smoke and all other areas have been searched, I'm okay with you informing Command that you have an all-clear on the building and unless otherwise directed are going to open some windows on the sixth floor to assist with salvage. Command will tell you if that's all right or if there are other assignments for you. Again, *communicate.*

Search and EMS/patient care

The relationship between Search and EMS groups is generally the product of low staffing. In a utopian world, the incident commander has staffing sufficient to conduct a search and, when a victim is located, a crew other than the search crew is available to bring the victim to the outside or another safe area. At that time the "rescue" group would turn the victim over to trained EMT or paramedic personnel to begin any required medical treatment.

Unfortunately, we don't operate in a utopian world. Real-world staffing prohibits such a luxury. Even with adequate or nearly adequate staffing, when a search crew finds a victim, they almost always remove the victim from the building. Not only do they remove the victim from the building, but they usually begin EMS or lifesaving efforts on the victim they've just brought out. I'm trying to remember in my career if I've ever noticed the same firefighters bring out a victim and then immediately return to the building to search for more (the operative word being "noticed," because a lot of things of that nature occur and for whatever reason I never see them). For the most part that just doesn't happen.

I have started asking firefighters in training sessions, when discussing search operations, if they can ever remember a firefighter bringing out a victim and then returning to the building to continue the search. No one has answered in the affirmative yet. That is not to say that it never happens, but it is the exception more than the rule.

It's a shame that in the richest country in the world staffing levels for the fire service are not commensurate with the need. Now more than ever, fire departments struggle to maintain an adequate number of personnel available to complete the tasks the community expects of us.

With regard to search and EMS, no one advocates removing a victim from a burning building only to leave them in the front yard without giving then the needed emergency medical attention. The decision on who will tend to located and removed victims is up to the incident commander. In most cases, until staffing improves or the utilization of automatic or mutual aid increases, patient care on victims found inside a burning structure will most probably remain with a search crew who finds them. Assuming that's the case, who will finish the search and what areas will be hit twice or not at all?

The Standard Method of Search

5

The most basic method of search is what I call the standard method. This is probably the type of search you were taught in recruit school. As I said earlier, prior to 1984 I didn't know there were other methods. My guess would be if you don't know any other method, you are probably using the standard method of search.

The Standard Method of Search Defined

The standard method is a group search whereby all members of the search team cover the same area together. Two or more firefighters enter a room, one or more goes to the left (left-handed search), and one or more goes to the right (right-handed search). Then, they meet somewhere in the middle of the room. After the room is searched, they make their way to the next area to be searched.

In this definition, note that two phrases are emphasized. *Group* is emphasized because a key to this search is the team effort in which the group stays together constantly throughout the search. *The same area together* is also important because with this method, only one area is searched at a time.

The officer participates in the search along with the crew. There are many officer responsibilities with this method of search. Among these are:

- **Crew safety.** This is the case as with all group/division assignments. Safety is the foremost responsibility of any group/division officer. Changing fire conditions/location and structural stability are among the safety points to be considered.

- **The search plan.** Ideally, while approaching the area to be searched, the officer is developing a plan for the order and path of the search.

- **The victim plan.** The officer must determine whether the crew is staffed sufficiently to split in the event that a victim is located? If not, a back-up plan must be developed.

- **Communication with command.** Keep the IC informed about the progress of the search and the crew's location (accountability), along with any other pertinent information.

- **Conducting the best possible search pattern while performing all of these tasks and more.**

Did you catch the last bullet? I'll repeat it: *conducting the best possible search pattern while performing all of these tasks and more.* That's a lot to do all at the same time.

Advantages to the Standard Method of Search

As I said earlier, the standard search is a group effort. The team enters together and remains together until the search is completed (unless a victim is located and staffing allows for two members to remove the victim while the search continues). There are several advantages to the standard search.

The search conforms with two-in/two–out

There are two parts to two-in/two-out. First is the condition that, prior to initial interior operations, a minimum of two members must be present in order to go into the structure to mount firefighting efforts. You must also have at least two members present outside on the scene to act in the event that the initial interior crew gets into trouble. The second part of two-in/two-out is the buddy system. The standard states that, while inside an IDLH atmosphere, members must remain in voice contact with each other.

This method conforms with the standard. Members of the search team enter and remain together inside the structure while conducting the search. With this search, each crew only searches one area at a time. If the size of the crew (at least four or more) allows them to be split into two teams, each team still searches one area at a time.

All members are responsible for a search area

Another key to standard search is that the officer participates in the actual search efforts along with the other crewmembers. So, for a four-person crew, four members are sweeping in corners and under beds. That means more members are searching at one time, and in theory, raising the likelihood of finding any savable victims.

Disadvantages of Standard Search

With standard search, safety is not foremost in anyone's mind. The first and foremost responsibility of any group/division officer is crew safety. Search is a complicated process and requires substantial focus to be done right. Even the best, most experienced officer can't effectively search for victims while also focusing on crew safety. Something will be shortchanged, and this is not the time or place to skimp on efforts.

The safety of the search crew entails several things:

- Assuring that crew integrity is maintained. It is the officer's responsibility to assure the crew remains together. This is very hard to do under the best of conditions with some firefighters. It is certainly more of a challenge if the officer is focused on sweeping in corners and under beds.

- Maintaining awareness of fire conditions. Rapid buildups of heat or fire conditions, along with rapid intensifying or lessening of smoke conditions, are among the factors the search officer should evaluate on a constant basis.

- Structural factors. There are very few "nevers" in the fire service. I have read a lot of textbooks that say "never do this" or "never do that." Anytime you hear the term "never" in a discussion of firefighting operations, red flags should go up. As an example, there are books that tell firefighters to "never" take your gloves off. Does that mean you can't remove your gloves while you are outside the building taking a break as well? What does that "never" mean to a three-month rookie?

- Over the last 20 years or so we have become totally encapsulated in bunker gear. This is truly a double-edged sword. On one hand, our gear does a great job at keeping heat away from us. On the surface, that's a good thing. But how do we tell if the fire environment is too hot? I can tell you from experience that if you are suddenly getting very uncomfortable due to heat, you are in trouble. With this encapsulation, there is no outlet or valve to tell when heat is climbing rapidly in the environment. In order to monitor changing heat conditions while encapsulated, you need to expose some heat indicator to the environment.

- Until SCBA or bunker coats are equipped with heat sensors, you must momentarily expose some skin in order to get a true sense of the current environment. Generally, you have to expose the palm of your hand. I must stress that this is a brief instance. Only expose your palm long enough to detect the heat in the atmosphere, and then pull your glove back over your palm. Until technology prevails, I know of no other way to accurately determine the atmosphere and still remain fully encapsulated, except for brief monitoring instances. If you momentarily expose your palm and experience high heat, then re-cover your palm and reevaluate the area you are searching. Could there actually be "viable" victims there?

- As it relates to search and structural stability as in the beginning of this bullet, how do you know the floor assembly you are crawling on is not being subjected to or involved in the fire, especially lightweight floor assemblies? There are two procedures that give good indications of the amount of fire and heat below you and how it is affecting the floor assembly stability.

- First—and least desirable—is using the weight of your body to determine the amount of sag or bounce in the assembly. As you enter any structure, one of your first actions as you crawl into the area involved in fire is to place your weight (bounce—don't jump) on the floor to feel the stability. If the floor seems spongy, sags, or bounces, the joist integrity may be affected by fire. This is especially true of lightweight construction. If the floor assembly appears weak, back out and go to plan B. If not, use that as a sampling of the assembly and proceed. Occasionally, repeat the same test to maintain awareness of the stability of the assembly.

- The other process to determine the amount of heat below is to feel the floor. This again must be done with bare skin. As you enter, especially in lightweight construction buildings, remove the base of your glove to expose your palm and feel the floor for warmth. An uncomfortably hot floor is not a safe sign; care must be taken from that point on. If the floor is not too hot (I realize that is a relative term), you and your crew can proceed, but it is your responsibility to occasionally test the floor as you proceed.

- Structural stability. It is the responsibility of the officer to assure safety in the structure while a search is being conducted. One of those aspects is maintaining awareness of the stability of the structure. This is very hard to do while sweeping under beds and in corners.

- Focus is concentrated solely on searching. There's the word focus again. I hope by now you're getting tired of me saying that. If you are, that means you're paying attention. There should be some kind of rule of thumb here that states that "the more smoke and heat present in a fire building, the less effective multitasking becomes." Incident command is built on focus, and every officer and firefighter on the scene should have a specific focus. If the search officer is focusing on crew safety and the search problem at hand, and the searcher's focus is on searching, then everything gets done to the best of the firefighters' abilities. Inside a burning building, multitasking cheats one thing or the other. What activity would you want to be cheated if firefighters were searching for you or your loved ones? I think the answer is evident. When conducting a standard search, all members are searching. This severely prohibits or curtails the ability of one searcher to focus on what an officer should be focused: safety.

- Only one area gets searched at a time. The standard search is a group search. All members of the search team remain in the same room or area and search. When conducting a search for victims during a fire, time is of the essence. Anything that can be done to shorten this time may mean the difference between life and death to the civilian.

- Continuity of search may not be maintained. Continuity of search is conducted in a logical and uninterrupted manner. This goes back to focus. If the officer's focus is on searching, then a well-thought-out search may not come to fruition. Something will be neglected or short-changed, with potentially tragic results. An officer should plan any search he or she will be involved in, but how many officers create a plan in the event a searcher locates a victim? Who will take the victim out? Who will continue the search in the area the victim is located? These are only a few of the elements necessary to maintain continuity of search.

Variations to the Standard Search

There are a few variations to the standard search. Out of the four types of search that I will cover in this book, the standard method of search is the least complicated. The way that you conduct any search is a product of training and necessity. Regardless of crew size, a standard search is conducted by sending half the crew in one direction in a room and the other half in the other direction (fig. 5–1). As they search, they sweep and cover their area of search until the two groups meet somewhere on the opposite end of the room or area being searched. At that point in time, there are several variations on what will be done next. Some departments teach that the searchers exit the room following one wall back to the door. Other departments teach that the two groups pass each other and rapidly conduct what some departments call a secondary search until they would make their way back to the door in which they entered. At that time, they move together as a crew to the next area to be searched.

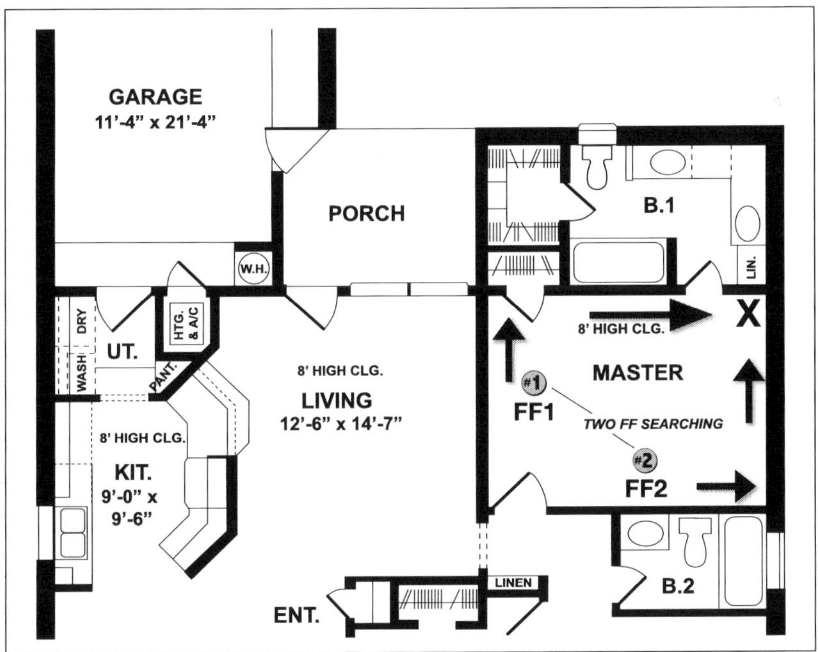

Fig. 5–1. Typical standard search. One firefighter searches to the left and the other searches to the right and they meet somewhere in the middle of the room.

Some departments conduct a standard search in an entirely different manner. The crew enters the area or room to be searched and one firefighter, generally the officer, follows the wall, maintaining constant contact with it. At the same time, the remainder of the crew crawls beside the officer, but more to the center of the room while maintaining contact with the officer on the wall. Some challenges to this method present themselves when furniture or other obstructions are encountered. Ideally this likely obstacle is handled through extensive training.

Some departments use the same process, but when staffing is short and only two members are conducting the search, the officer maintains contact with a wall and the other firefighter moves laterally toward the center of the room and then back to the officer. Then they both take a few crawls forward and the process is repeated. This ensures that the entire area including the center of the room is covered.

One last variation on the standard search consists of the crew entering and the officer maintaining contact with the wall, but the searcher moves much more toward the center of the room. The searcher maintains contact with the officer utilizing a short, 3–5-ft. piece of webbing or a tool such as a Halligan or an axe.

The Standard Search in Detail

The size of the crew

The minimum size of any crew conducting the search should be two firefighters with at least one of them being an officer. Not only is this common sense, but state and federal standards dictate that the minimum size of a firefighting crew working inside a building is two firefighters.

Experience has taught me that there is a maximum number of searchers for conducting an interior search in zero visibility. That number is five firefighters, including the officer. Two factors play into this: span of control and span of communications. Regarding span of control, one officer is hard-pressed to maintain accountability (location and function) of more than four searchers at one time. It is equally hard for one officer to communicate effectively with more than four firefighters while managing a search in zero visibility, especially with the additional constraint of the SCBA facepiece in operation. The maximum number of searchers varies with the type of search being conducted. In relation to the standard search, I believe the number is anywhere from two to four members, including the officer.

Entering the building and moving to the area to be searched

As stated in previous chapters, the officer should be developing a plan while moving to the building or area to be searched. When utilizing the standard search, the officer should lead the crew into the building and head to the selected area to start search efforts. If visibility allows, the officer can visually locate the stairway or other target areas. In instances of zero visibility, the officer can follow a wall to the initial destination. In subsequent chapters I will discuss where to search first and where to search last. All members assigned to the search team should remember their primary exit point (where they entered the building) and closest emergency exit, such as the last window passed.

Conducting the search

Upon locating the room or area to be searched, the officer should split the search crew into two groups, then direct one group to go to the left and the other group to the right (if that is the type of standard search they will be conducting). The officer, along with the remainder of the crew, then searches the left room or area until search is complete, at which time they move onto the next room or area to be searched.

Method 1: Splitting the search team for left-hand/right-hand search. Moving into the room to be searched, the officer splits the crew in the most productive and safest manner for the number of crew members. If the crew is simply the officer and another firefighter, the officer will direct the firefighter to go to the left (left-handed) and the officer will search to the right (right-handed) in the same room or area. With three members, two can go in one direction and one firefighter or the officer alone will go the other direction. Later I will get into the specifics of how to conduct a search using "your wall" as your point of reference to your way out of the search area (fig. 5–2).

As these searchers or groups of searchers move about the room in their direction of search, I expect them to maintain a point of reference with "their wall," then move or sweep laterally toward the center of the room using side crawls, and then crawl again, laterally, back to their wall. It may sound confusing at first, but as you read on it should become clearer.

Once the two groups meet somewhere on the far wall, departments generally teach one of two procedures to move back to the door or point of entry. The first procedure is simply to have the two groups merge together as one and follow one wall back to the door. The second procedure is to have the two groups pass each other and continue conducting what some may call a "secondary search" in the area previously searched by the other group.

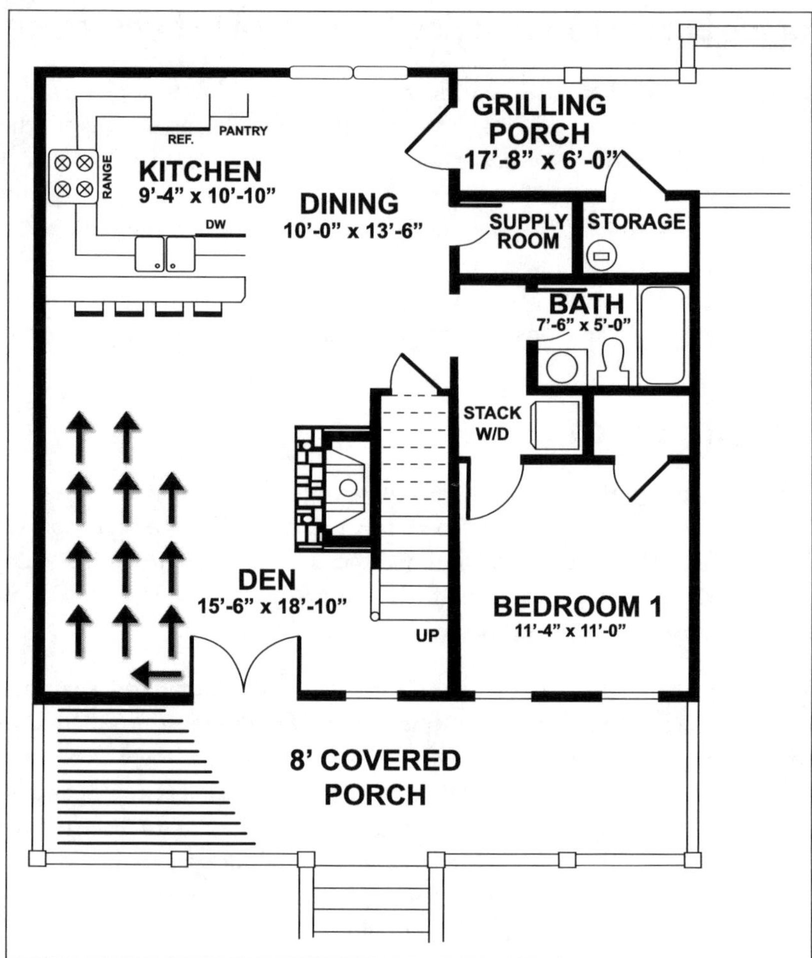

Fig. 5–2. A searcher following the "searcher's wall." As long as awareness of the wall is maintained with the same shoulder (in this case, the left shoulder is *always* parallel with the wall) the searcher knows where he or she is in the room and related to the exit.

Method 2: Searching the area as a single group with one firefighter on the wall. This is the second variation to the standard method of search. In this method the crew enters the room to be searched and the officer informs the crew whether they're doing a left- or a right-handed search. At that time the officer will enter the room and maintain contact with the selected search wall. If doing a left-handed search, the officer's left shoulder or arm will *always* stay in contact with the wall. As the officer moves along the walls around the room, the rest of the crew crawls parallel with the officer toward the center of the room. If there is only one firefighter with the officer, that firefighter should, every two or three crawls forward, move laterally toward the center of the room. The firefighter should sweep with the right arm toward the center of the room, then crawl laterally in the reverse direction back to the officer. At this time they will both move forward two or three crawls. They then repeat the same process of the firefighter moving out laterally toward the center of the room. If there are more than two members in the crew, the remainder of the crew can spread out toward the center of the room while maintaining lateral contact with each other.

There is a variation to this method, wherein the officer maintains contact with a wall as in the previous paragraph, but the firefighter or firefighters use webbing or a tool to increase the distance between the officer on the wall and the center of the room.

Standard Search and Other Groups/Divisions

Firefighting is a group or team effort. Everyone has a specific task (focus) and, as long as everyone completes his or her task, the job gets done safely and effectively. This is not, however, like an assembly line at an automotive plant where stations are fixed, visibility is excellent, and working conditions are, for the most part, ideal. In our line of work, groups and divisions must work not only independently, completing their specific evolutions, but also in the presence or in close proximity to other divisions or groups, which are often times obscured from one another due to smoke conditions. It is important that each group or division working inside a smoke-filled atmosphere has a basic idea of what other groups and divisions are doing.

It is the officer's responsibility to assure that two groups or crews working in the same area do so without hampering or limiting the other group's ability to complete their evolution. Communication is the key and this is one of the main reasons why I advocate that the officer in charge of any group or division is not a working supervisor, but more of a "supervising supervisor."

The officer conducting a standard search and the attack officer

The officer assigned to search and an officer assigned to attack have different focuses. The search officer's focus is search crew safety and assuring a rapid, thorough search is conducted in any area where savable victims could be located. The attack officer is focused on attack crew safety while they're busy putting out all the fire in their assigned area.

As long as the attack group and search crew operate in different areas, there should be little cause for concern. It is incumbent upon both officers to be aware of the other group's location. In my opinion, it is more important that the attack officer knows where the search crew is than the other way around. This is especially true if the backup group has not yet been assigned. The problem arises when the search officer believes there are savable victims in the area where the attack group is operating. This would be the case in a fire involving a bed or in a bedroom closet with zero-visibility conditions. The attack officer's responsibility is to darken the fire. If the officer believes the fire has been contained to that room of origin, attack would begin that darkening process by removing the mattress or bedding material. This is very hard to do when a search crew is trying to conduct a primary search in the same room.

As said before, I do not advocate an attack group conducting a search or a search group mounting a fire attack. Focus will be lost somewhere. However, as always, "never say never." This is one of the few times I do advocate that the attack group immediately, upon darkening down a fire in a bedroom in an occupied house, conduct a very rapid search in that bedroom to determine if any victims are present in that room or area. Immediately upon search completion I expect two things to happen. One, the attack officer informs command and the search officer via radio that attack has conducted a search in the fire room. Two, the officer immediately returns focus to attack officer duties.

The officer conducting a standard search and the vent officer

Often the search officer will be in a position to provide information to the vent officer concerning the effectiveness of the ventilation efforts. It never hurts for the search officer to tell the vent group officer how visibility and conditions are changing in the area being searched. Going back to the word focus, I don't believe searchers should open up windows for the purpose of ventilation while conducting search.

I say this for two reasons. One, it takes precious time to open a window and clear the window completely. I don't advocate opening a window for the purpose of ventilation without clearing the window of all glass in the event of an emergency and the needed utilization of that window for egress. This takes away precious time from the search.

Two, I don't advocate searchers opening up windows, because it may not be in the ventilation plan being conducted by the vent group officer. This is especially true if PPV is being used. Not only may it negatively affect the efforts of positive pressure ventilation, but there is a physical tendency to draw hot gases toward the opening being created by the searcher. That can have negative effects on the search effort and, certainly, may deteriorate conditions for un-located victims.

The officer conducting a standard search and the backup officer

When staffing becomes available, a backup group should be assigned to shadow the search crew. In the absence of the assigning of search, the backup group should shadow the attack group. However, in residential occupancies we generally send the search group in without a hose line. This will be discussed later. In that case, backup should, as a priority, shadow the search group if it is assigned.

The backup line should be positioned so as to not interfere with the search. There should be communication between the search group officer and the backup group officer to assure that the search is being conducted in the safest, most rapid manner possible.

Team Search

6

Of the three basic forms of search that I am aware of, team search is the most complicated, least understood, and least trained for. Standard and oriented searches can be conducted in more occupancy types than team search. In fact, team search has firefighter rescue as its primary application. An important application, to be sure! However, not too many other occasions call for a team or rope search.

Team Search Defined

Team search is defined as a method using a rope or other flexible and easily deployable tool to provide firefighters an awareness of their route of egress. This rope plays out as the team enters, regardless of visibility conditions, and is anchored to a stationary object when the direction of travel is changed. It is there so that, if conditions worsen, firefighters can leave the building. The rope can also be used as a tool to assist another crew that needs to locate the search team.

As I said at the onset, I believe team search is very complicated and time consuming. It is not a search generally thought of when civilian lives are at risk. It is generally used to locate and remove injured, lost, or trapped firefighters. I have developed a list of applications where I believe team search should or could be used:

- As a tool for rapid intervention teams to locate and remove injured, lost, or trapped firefighters in large areas such as warehouses and factories

- As a tool to locate fire in a warehouse with cold smoke due to the activation of automatic sprinklers

- As a tool to locate the most appropriate doors and access points in warehouses and other large area occupancies with cold smoke due to the activation of automatic sprinklers

- As a tool to assist in locating fire in larger strip mall occupancies

- Any time crews need to enter and reconnaissance inside a large structure or occupancy where there is a chance that visibility conditions may become obscure

There are many variations of the use of rope as a search tool as in team search. I believe they have evolved from the same basic principles. I'll describe what I believe they all have in common.

Most firefighters conducting team search use rope as the focal point of the search. By that, I mean that rope is the tool at the very heart of the search. Today ⅜-in. or ½-in. Kevlar or kern mantle rope is used for team search. Some departments also carry thinner, steel-braided rope if high heat conditions are anticipated. Almost every department that uses rope to search large areas carries that rope in some type of bag that allows for ease of storage and deployment.

Most use 200-ft. or longer ropes. The maximum for a single rope is generally 400 ft. The weight of the rope and bag, as well as the ease and time required for deployment prohibit ropes being much longer than 400 ft.

Most place or tie knots in equally-spaced intervals, usually every 20 ft. (fig. 6–1). The distance between the knots and the point of entry is often indicated by the number of knots. For example, three knots indicate that you are 60 ft. (20×3 ft.) into the structure. Some departments also place a ring next to the knots to provide a place to anchor or lock onto the rope, and also to serve as an indicator as to the direction you are traveling on the rope (fig. 6–2). Knot then ring in or ring then knot out!

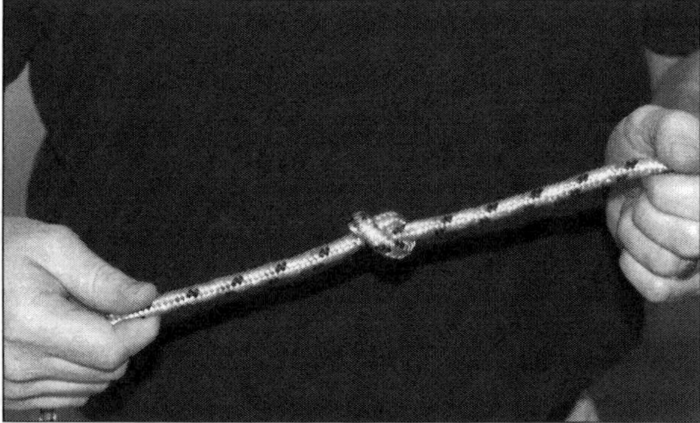

Fig. 6–1. An example of the knots tied in search rope

Fig. 6–2. A ring tied into the search rope for anchoring onto the rope or to serve as a direction indicator

Advantages of Team Search

Even as complicated and at times cumbersome as team search can be, it does have distinct advantages.

- **The Search conforms with two-in/two-out.** Team search is called team search because you enter and remain as a team throughout the search. Hence, it easily conforms with two-in/two-out, assuming there is another two-member team outside as backup for the two searchers.

- **Orientation with means of egress.** If done correctly, team search provides a constant path to the outside or a safe area for the search team.

- **Provides a safe, methodical way of searching large open areas.** Utilizing the search rope, members can play off of the search rope using individual tag lines to conduct searches of large, open areas and still maintain an awareness of egress if necessary.

- **Easy location.** As with pulling and stretching a hose line into the interior of the house, team search and the utilization of the rope allows crews entering to locate the search crew in a rapid manner.

Disadvantages of Team Search

As with almost everything in the fire service, there are advantages and disadvantages to even the best techniques. Firefighting is not a "one-tool-fits-all" profession. That's part of what makes this job so challenging and the training function in any department so vital. We must train on all of the tools available so that when the situation requires a specific tool to be pulled off the rig, we are familiar with its operation and limitations.

- Sole reliance on the path of egress is placed upon the rope. As with many evolutions utilized in the fire service, focus diminishes with multi-tasking. This is especially true as smoke conditions increase. The anchorman's main focus is on playing out the rope. The officer's main focus is the safety of the crew and the search itself. Searchers should focus on searching. The ability of a crew to find their egress in the event that conditions worsen is made easier if the search rope's integrity is maintained. However, if the integrity of the search rope is not maintained for any reason, the likelihood of crewmembers locating their exit in zero visibility after focusing on different aspects of the search is greatly reduced.

- Crew members seem to spend more time on the physical aspects of utilizing the rope and tag lines than they do searching. This factor is greatly increased as visibility decreases. This is another major concern that I have with team search. If used only as a tool to provide initial crews a means of egress in large areas in the event that present visibility conditions worsen, and not as a tool to locate lost trapped or injured firefighters or civilians, this is an excellent tool. However, if used as a primary tool in IDLH conditions to locate endangered firefighters or civilians, team search is not a rapid method of locating victims. If done correctly team search is time consuming. That fact reduces the survivability of victims.

- Crew safety may not be the primary focus of the officer. Because of the complexities of team search, safety may not be foremost in the mind of the officer or team leader. Increased fire conditions reduce one's ability to multi-task effectively, and it is in times of increased fire conditions where safety should take precedence over almost everything else.

- Continuity of search may not be maintained. The biggest stumbling block to maintaining continuity during a search is when we locate a victim. It is at that time when our focus turns from searching, which hopefully is organized and planned, to rescue, which is normally unorganized and chaotic. Once the victim is taken to a safe place, the likelihood that the same crew will re-enter and pick up where they left off becomes remote for a myriad of reasons.

Team Search in Detail

As with the standard search, there are many variations to team search. The type of rope used varies from department to department, as well as the distance between knots tied in the rope and many other details. Having looked at several variations in team search, I have come up with what I believe to be a very basic or generic form of team search.

Team size

The size of the crew using team search is generally dependant on the size of the crew they responds on an apparatus. Statistically that would be three to four members. What I describe below will focus around a four-member crew, including the officer. As with any type of search, any more than one officer and four firefighters stretches the span of control and span of communication between the officer and the crew members. It is extremely difficult for one officer to control the actions and communicate effectively with any more than four firefighters at a time. This is certainly the case in IDLH conditions.

Team members

The anchor. The firefighter in front of the team with the bag is called the anchor person or the "anchor." The anchor takes direction from the company officer who is the second person on the line. The anchor has four responsibilities:

- Secure the search line to an anchor point outside the building or in a non-hazardous area. The anchor should attempt to tie off outside the building at least 10 ft. from the door or entrance (fig. 6–3).

- Feed the line out of the bag as you move throughout the structure.

- Call off the number of knots as they appear on the rope in order to inform the remainder of the crew as to how far into the building the team is.

- Maintain tension in the search rope. The search rope should be positioned approximately 18–20 in. from the floor, making it easier to locate.

Fig. 6–3. The anchor connecting the rope to a fixed object outside the building

The officer. The next person on the line is the officer. The officer has several duties:

- Responsible for the safety of the team

- Responsible for developing a plan of search

- Responsible for checking the air supply of each member prior to entering the building. The officer should gauge the working time inside the building based on the member who has the lowest amount of air in the bottle and each member's general air consumption rate. Air management is a relatively new concept in the fire service. Fire Engineering Books has an outstanding text on air management by a group of authors from Seattle called *Air Management for the Fire Service*.

- If the search team has a thermal imaging camera (TIC), then the officer is responsible for utilizing TIC.

- Conducting the search in a logical, efficient manner.

- Communicating with command.

Searchers. The next firefighters on the line are called searchers. Their task will be to:

- Follow the anchor and officer into the building until the officer directs them to conduct a search in the area. This search will be conducted in one of several ways, depending upon fire conditions and training.

Conducting a search

I would be hard-pressed to come up with a reason why team search would be used in a residential occupancy. Residences do not lend themselves to the application of team search. Team search has as its major application commercial occupancies that are large and wide open. Because of this, I will focus on team search as it takes place in a large commercial space.

Entering the structure. The search officer will determine where the team will enter the building. There are few factors that need to be determined when making this decision. Among them are the location and the amount of fire in the building, as well as what the search team's mission will be. Are you entering to a) locate downed, lost, injured, or trapped firefighters or civilians, b) search for the location of the fire in the event of sprinkler activation with smoke hanging near the ceiling, or c) locate the attack crew and the area where firefighters or civilians are believed to be?

The search rope should be anchored approximately 10 ft. outside of the building. This is done to assist in identifying the location where the search team entered the building and to give them room to gather outside the building in a safe area in the event that they had to leave due to heat or fire conditions. The search officer will then lead them into the building and to the first area to the searched. The entire search steam stays on the right side of the rope, maintaining contact with the rope using their left hand. This remains constant throughout the entire search, including exiting the building. As they enter the area to be searched, the search team officer should let the anchor move ahead and the team should stop at the entrance while the officer scans the area with the TIC. The crew members should stand shoulder to shoulder while this is being done and the officer should scan both high and low. The officer should also let the other members of the team scan the area with the TIC prior to moving in.

Moving through the building. After scanning the area with the TIC, the officer determines the direction of travel and informs command ("Search to command: we're heading north into the building toward side C"). The anchor then moves forward toward the C side of the building and the rest of the crew follows. Using hands to feel the knots play out of the bag, the anchor informs the crew of the number of knots that have just passed through, which indicates how far they are into the building ("two knots, 40 feet in").

When the officer wants to change direction of travel, the officer informs the anchor, who wraps the rope around a substantial object 18 to 20 in. off the ground to maintain tension on the line, and then the anchor moves toward a new direction of travel. At this time the officer will inform command of the change in direction, approximately how far they are into the building, and their new direction of travel ("Search to command, we are now heading west toward side D and we are between 40 and 60 feet into the building"). Also, the officer should check the air supply of the team members.

This would be a good place to discuss when it is safe to walk in the building. My rule of thumb (which I stole from Tom Brennan) is, if you can see the floor then you can walk. The converse is also true. If you cannot see the floor, then you should crawl.

Searching off the search rope. If the team is searching for victims, there might come a time when the officer wants the searchers to branch out laterally off the search rope (fig. 6–4). There are several ways that this can be accomplished. One method is for the officer to remain on the search line and have the searchers crawl laterally, still facing in the original direction of travel, a specific number of side-crawls away from the rope. Once they reach a predetermined number, they move laterally back to the officer and the rope.

Fig. 6–4. Searchers searching out away from search rope

Another method is to have one searcher connect to the search line using the ring next to the knots. The officer then instructs the searcher to do a 360-degree search pattern. The searcher will snap into the ring and move ahead to the 12:00 position on the line, playing out his tagline to its end. This will move him to the next group of knots if he is using a 20-ft. tagline (fig. 6–5). The searcher will then start to move to the left or the right off the search line, creating a circle. The searcher should inform the officer of the intended search direction (left or right). When the searcher has traveled 180 degrees back to the search line, he should inform the officer that he is halfway done and now moving again along the opposite side of the search line and back to a starting position. Using this method, two searchers can safely cover a significant amount of ground.

Fig. 6–5. Searchers moving in semi-circles back to officer's position

A third method is accomplished by having two searchers snap off to the same ring and each complete a half-circle search in an area opposite the search line from one another. The officer remains as an oriented member at the point where the firefighters snapped to the ring. When conducting this method, the firefighters again snap to the ring and then move to the 12:00 position, then start search out away from the search line. When they meet at the other end of the search line, they crawl back up the line to where the search officer is positioned.

Some departments have the searchers simply gather their tag line up in their hands as they move back toward the officer. Other departments have them wrap the tag line around their wrist in a circular motion while moving back to the officer. This way the line can be played out again rapidly by simply letting it come off the searcher's wrist.

Exiting the building. If the search proved negative and the officer determines it's time to exit the building, then the anchor should drop the search bag and the officer should move to the last position on a line. This is done to assist with crew accountability. If the line was recently anchored, the crew can proceed out of the building by following the line back toward the point of entrance. If the search line was not recently anchored, the officer should again move to the back or the rear position on the line and drop the search rope to the ground while maintaining tension on the line (fig. 6–6).

Fig. 6–6. Crew preparing to exit

The crew then moves to the other side of the line, maintaining contact with their left hands. Then the officer raises the line and they all head back toward the entrance. Every 10 ft. or so, the crew should pause while the officer moves up to their position and then places tension back on the line so the crew can again move forward. This process repeats itself until the crew comes up to the last anchor point where the line was wrapped. At that time the crew can exit together and the officer need not maintain tension on the line anymore.

There are many variations to the application of team search. I have laid out a basic principle for application that you can modify to fit your own evolutions. Team search is a complicated method of search, but I believe I have provided a very standard and safe application. Later in this text we will get into specific occupancy types and, where applicable, I will illustrate how this method of search can be used as safely as possible.

Limitations to team search

Team search is a tool that has specific applications to the fire service. However, of the three (or four) main types of search, I believe it is the most limited in its application.

It is generally not conducted in residential occupancies. These buildings are too small and cluttered to make this an effective method of search. Even large multi-family residential occupancies are prohibitive to team search. The reason team search is a method at all is because the rope provides orientation to the area of egress in large, wide open buildings. In apartment buildings, the hallway provides us with that orientation. Maintaining an awareness of the number of doors you have passed down a hallway will be sufficient in locating the egress in apartment buildings.

Team search is very time consuming. If done correctly and safely and in IDLH conditions, much time is required in playing out and maintaining orientation with the search line utilizing tag lines. This task becomes even more challenging when firefighters use their gloves. It is for this reason that I do not consider team search as a primary "go to" evolution in life or death situations. I realize many fire departments count on this form of search as a primary tool for RIT and firefighter rescue. I also do not advocate that fire crews enter large-area buildings without some means of remaining oriented to their point of egress. If a firefighter rescue is the only reason we are using team search then, in my opinion, it is too slow and cumbersome to be effective.

The safest and most effective use of team search is in cold smoke conditions with sprinkler activation and large warehouses and other commercial occupancies where visibility is not a current factor. This is an excellent tool when entering such occupancies to look for the seat of a fire or to find alternate areas of ingress and egress. The rope can easily be played out in ideal conditions with excellent visibility.

A Few Last Words on Team Search

If you are contemplating conducting a search in a large area, you must use something to remain oriented with your entry point. You can use a rope. You can use a hose. You can use steel cable. Whatever it is, use something. Whatever that something is, do it the same way every time. Train with it, practice with it, and do it that same way when called upon to do it at a real fire. There's a psychological principle that I will refer to several times in this text. That principle is, "When people are put in stressful situations, they revert to what is customary and routine."

Actually using a search rope in zero visibility to locate an injured, lost, or trapped firefighter or civilian will be an extremely stressful situation. This is not the time to conduct an evolution that you haven't drilled on for several years. This is the exact time when your mind and body will most effectively do what it is accustomed to doing. It is for this reason that, if you're actually going to consider conducting a team search under stressful conditions, then it has to be as close to second nature as brushing your teeth in the morning or putting on your protective hood en route to a fire.

"Can we realistically do what we intend to do?"

When considering team search, the incident commander or the search officer should ask one very specific question. That question is: how long will it take a four-person crew in zero visibility to search a 120,000 sq. ft. warehouse? (That's not a big warehouse.) Proper air management techniques tell us that we should never be allowed to operate inside a building when our low air alarm is sounding. The officer needs to be aware of each crew member's air supply at all times. There are those who say we can get lax in residential occupancies where there is a window or a door every 15 or 20 ft. Most of the firefighters who believe that never have felt the pain of losing a brother or sister inside one of those "bread and butter" fires. To even contemplate operating inside one of these large warehouses or other occupancies such as a box store and not be cognizant of air consumption and quantity is playing Russian roulette with not only yourself, but your crew.

Let's do a little math. With a 30-minute SCBA bottle, the average firefighter has between 15 and 22 minutes of actual work time before the low air alarm begins to sound, or less depending on actual consumption. The firefighters should have exited prior to the activation of the alarm. At the maximum (22 minutes), that will give the search crew 11 minutes to get in and operate and 11 minutes to exit. That means the crew would have to cover approximately 11,000 sq. ft. every 60 seconds.

Think of it this way: I would guess that the average firefighter's home today is somewhere between 2,500 and 3,000 sq. ft. Do you think you and your crew can effectively cover what amounts to more than three houses in 60 seconds?

I sincerely doubt it.

The Oriented Search

7

This is the last of the three main types of search with which I'm familiar. Chapter 8 discusses one last method of search, but I do not consider it one of the major methods.

This chapter will focus on the *oriented search*. In my opinion, the oriented search is the most effective and safest search method for the majority of fires to which we respond. As you'll soon see, if done correctly, there are many components to this method of search.

I first learned of this method in 1984 as a lieutenant for the Toledo, Ohio, fire department. Our department had just hired 64 new fire recruits and I was one of 12 instructors removed from field operations and detailed to the training bureau. I was chosen to teach SCBA and search operations to the recruits. In preparing my lesson plan, it dawned on me how little I actually knew about conducting searches.

In my 1975 recruit class, we spent approximately half an afternoon on conducting searches. After a lot of research through magazines and textbooks (there was no Internet in 1984) and discussing search methods with other instructors, I stumbled on an article in a trade magazine titled *Surviving the Search and Rescue* (I have since tried to locate that article to give credit to the authors, but to no avail). The premise behind the article was that search is an inherently dangerous evolution that can be made safer by altering the traditional duty of the officer assigned to search. In those days, the officer participated in the search as in the standard method of search I discussed in chapter 5. This article examined the possibility of allowing the search officer to focus on crew safety and the search plan by removing the officer's responsibility of actually conducting the search.

I read and reread the article and then developed the oriented search based on the premise of that article. From there it has evolved into an accepted and internationally used method of search.

The Oriented Search Defined

The oriented method of search is defined as using an officer or "oriented position," and one or more searchers conducting a one-firefighter search evolution. The key to this method is that the officer is not actually searching—that is, not sweeping under beds or in corners and closets. The oriented position has a specific list of duties to focus on while the searchers are conducting one-firefighter search patterns. This allows focus to be split into two separate and distinct places:

- The safety of the crew

- The actual search

Advantages of the Oriented Search

As with any application or evolution that we use in the fire service, different methods of search have their own distinct advantages and disadvantages. There are times that specific evolutions should and should not be used.

- **The safety of the crew is maintained in all times.** The most distinct advantage of the oriented method is that it removes the officer from the role of physically searching for victims, leaving the primary duties as ensuring the safety of the officer and the crew. Search is an inherently dangerous evolution. This method allows one member to focus on the safety of the crew members who are conducting the search. In doing so, the oriented officer should be focused on three main conditions:

    ▷ **Fire conditions in the area of search.** Increased heat and rapid changes in smoke conditions are two major factors that occur when fire conditions are increasing rapidly.

    ▷ **Structural stability in the area of search.** The oriented officer should constantly be aware of changes in the structural integrity of the floor assembly below and the ceiling area directly overhead. The floor area below should be rapidly evaluated for increases in heat. Sagging or sponginess in the floor assembly should be evaluated as the oriented firefighter moves about the areas being searched.

    ▷ **The advancement of fire upon the search crew.** This is a specific concern when searching the second floor or other areas above the main body of fire. If a stairway was used to access the area being searched, then the oriented firefighter should occasionally check the stairway to ensure fire is not advancing on the search crew. The oriented firefighter should always be cognizant of heat and hot gases traveling from areas of high pressure toward areas of low pressure. When a searcher opens the door to a room to be searched, an area of low pressure may be created. If for some reason the window to this room is also opened or broken, an area of low pressure is created, which increases the likelihood that heat and hot gases will be drawn toward that low-pressure area outside the window. It is for this reason the oriented officer should occasionally check the stairway used to ensure that egress is available if necessary.

- **Searchers are allowed to focus on conducting search patterns in their assigned areas.** Imagine the stress of conducting a search at a real fire where the likelihood of victims being trapped inside exists. I have conducted many searches under these conditions prior to implementation of the oriented search. Now imagine conducting a search knowing an oriented firefighter is immediately outside the door to the room you're searching and is focused on your safety. That person constantly maintains awareness of fire conditions in the area. Much of the stress associated with conducting a search is eliminated with this realization. The searcher now has only one focus, and that is to locate endangered civilians. In essence, the searcher can move as fast as effectively possible to conduct the search. No time needs to be wasted on assessing changing heat or fire conditions in the area of search. Additionally, if the firefighter conducting the search becomes disoriented or injured while conducting the search, the oriented officer is only a yell away.

- **Searches are conducted faster.** In tests conducted in North Carolina in the early 1980s and in Toledo in the mid to late 1980s, it was proven that even after removing the officer from physically conducting the search, searches were still conducted faster using the oriented method of search. If taught correctly, one searcher can cover a room in a residence faster and more efficiently than two searchers using the standard method of search. The main reason is that the searcher no longer needs to be aware of personal safety while conducting the search. Hence, focus can be placed on moving through the area in a logical manner as rapidly as possible.

- **Search continuity is maintained.** Previously in this text I discussed continuity of search. Continuity entails knowing where we have searched previously, where we are searching now, where we will search next, and when the entire area has been covered, *even if we find and remove the victims.* This is a daunting task by itself, but failure to do so has probably contributed to more civilian deaths than the fire service would want to admit. Failure to maintain continuity usually occurs when we bring out a single victim without continuing on the search. Often we lose track and can't ensure that areas aren't missed or searched multiple times needlessly. The oriented officer should always be aware of where searchers were when the victim was found. The officer should bring either the same searcher back to the previous position or indicate to a new searcher where the victim was found and where the previous searcher left off. In doing so, continuity of search is maintained.

Disadvantages of the Oriented Method of Search

Based on countless training sessions using the oriented method, I have only come up with one disadvantage of this type of search:

- **It requires a great deal of concentration from the oriented officer.** If you think of all the responsibilities placed on the oriented position, you realize that many things will go through the search officer's mind while managing the search. Ensuring that a search is conducted in a logical manner and that the safety of the entire crew is maintained takes effort and concentration. This responsibility is not to be taken lightly. If done correctly, this is not a time for the oriented officer to rest in a doorway while the searcher does the actual physical search. Quite the contrary, there is a litany of tasks to be done while the searcher is searching.

The Oriented Method of Search in Detail

The oriented search is based on focus. The oriented officer has many areas in which focus will be required. While ensuring both personal safety and that of the crew is the primary responsibility, the oriented officer also has many search-related tasks to focus on.

The responsibilities of the oriented officer

Crew safety is far and away the officer's primary responsibility. As stated above, it entails several responsibilities relating to safety.

- **Monitor air supply.** If your department practices air management, then one responsibility for the oriented officer will be the air consumption rates and air supplies of every member on the search team.

- **Monitor fire conditions.** Fire conditions need to be constantly monitored, not only in the current area of search but in previously search areas and those to be searched next.

- **Monitor structural conditions.** Never say never! I have read in textbooks the statement "never take your gloves off." While this statement probably holds true for the firefighter on the nozzle, or with a carbide-tipped chain saw in hand, or conducting the search, this "never" does not apply to the oriented member. In many cases the officer or oriented member is positioned with the crew in an area remote from the protection of a hoseline, or in certain life-or-death situations above the fire.

We have totally encapsulated ourselves as firefighters. This may be a good thing on its surface, and I'm sure that burn injuries to firefighters have been reduced significantly as a result. Our bunker gear is designed to reduce our ability to feel the heat in the operating environment. If you suddenly become very hot in your bunker gear—hot enough that you are becoming uncomfortable and feel the need to physically move to eliminate the heat—you are probably in trouble. Ask the manufacturers of bunker gear or sales staff. I have, and they admit that if you are getting uncomfortably hot in your bunker gear, you are in trouble. My question then becomes, how can the oriented officer tell if the operating environment is becoming too hot too quickly? My answer is that we should use our ability to detect heat with our skin. Absent future technology, I can only offer the following.

To safely and rapidly do this, there are two options:

1) Quickly pull your Nomex hood from your cheek to detect heat, or

2) Pull your glove down to the point where the palm of your hand is exposed to heat. Either of these options should only be conducted for an instant—if excessive heat is detected, the search process should be rapidly re-evaluated for that specific area.

Regarding structural conditions, I advocate placing the palm of the hand on the floor occasionally to see if there is heat buildup in the floor assembly below. Waiting for this heat to be detected by conduction through bunker gear may prove to be too little, too late. Additionally, sagging or sponginess in the floor assembly should be occasionally monitored.

- **Ensure there is a safe means of egress.** The oriented officer should occasionally ensure that the way out of the building is safe from advancing fire. Generally this only requires moving back to the top of the stairway on the second floor landing or back toward your oriented point on the first floor if that's where you're operating.

- **Create a search plan.** This entails developing a plan of where to enter the building, where to start the search, where to proceed, and where to finish the search in a logical manner.

- **Know the location of crew members.** This is one reason that span of control needs to be narrowed for search. In the early 1990s in Toledo, we conducted training and found that one officer cannot effectively control more than three searchers. At that time Toledo had six five-person engine companies. When they trained on the oriented search, the officer had difficulty tracking the locations of four searchers. For us, an officer and three searchers worked best. Not only does span of control play a part, but span of communications prohibits a ratio of more than one oriented member per three firefighters or searchers. It is the responsibility of the oriented officer to know where searchers are at all times. This will be discussed in greater detail later.

- **Maintain awareness of search continuity.** This entails ensuring that the search is proceeding in a logical and interrupted manner and that all areas where savable victims could be located are searched at least once.

There are a lot of other tasks and responsibilities for the oriented officer. I will cover these in subsequent chapters when applicable.

The responsibilities of the searcher

The searcher has one focus, and that is to search. The oriented search is based on one-firefighter searches. The seriousness and the safety concerns of conducting one-firefighter searches is not to be taken lightly. However, the oriented search conforms with the two in/two out rule. As such, the oriented officer must be in voice contact with every searcher assigned to the crew at all times. The oriented search was developed for residential occupancies and, for the most part, lends itself well to this application. It has also been modified to work well in apartment buildings.

One-Firefighter Searches

One-firefighter searches are exactly like they sound. One firefighter enters an area or room to be searched, alone, with the intent of covering the room in a logical manner to locate victims.

Left- or right-handed search

Generally, left-handed persons tend to do left-handed searches. Right-handed persons tend to do right-handed searches. When doing a left-handed search, the searcher enters the room and moves to the left, maintaining a constant awareness of his or her search wall. I will also refer to this as "the oriented wall."

Throughout this text I will use the term "your wall." That refers to the wall that represents your awareness of egress, specifically with how it is oriented to the door used to enter the room. When doing a left-handed search, you should crawl parallel to the wall on your left. Your left shoulder and your wall provide you with an awareness of where you are in the room at all times. When moving throughout the room, your body should always be parallel with your wall (fig. 7–1). Doing so allows you to move forward through the room while always maintaining an awareness of the path to safety. Walls don't move. If a wall moves while you're inside a building doing a search, you have bigger problems than the search you are conducting.

Fig. 7–1. Maintain an awareness of "your wall."

By doing a left-handed search and maintaining contact with your wall, you can successfully cover the parameter of a room. However, victims have been known to be located in areas other than the perimeter of a room. In order to cover the room completely, you need to somehow move toward the center of the room while still maintaining orientation with your wall.

There are very few "nevers" is in the fire service, but this is one of them. *Never* turn perpendicular to your wall to crawl toward the center of the room (fig. 7–2). In zero visibility it is very hard to crawl, turn around exactly 180 degrees, then go back to the point on the wall from which you left. In order to do this as safely as possible, crawl sideways moving laterally away from your wall. Doing so always keeps the shoulder of your search—either your left shoulder or your right shoulder, depending upon your direction of search—closest to your wall.

In my opinion, the most effective search pattern is to enter the room to be searched orienting your shoulder with "your wall," then moving one or two crawls along the wall. To search into the center of the room, crawl laterally three or four side crawls toward the center of the room (fig. 7–3). Sweep once with your hand toward the center of the room and then move back to your wall. If you do this correctly, your head will always be pointed in the same direction that you are moving along the wall. When you get back to your wall, crawl forward two or three steps and then repeat the same process, moving laterally toward the center of the room.

Fig. 7–2. Never turn perpendicular to your wall to crawl toward the center of the room.

Fig. 7–3. Always keep the shoulder of your direction of search (left hand search = left shoulder, right-hand search = right shoulder) closest to "your wall."

If you are conducting a left-handed search and come to a corner in the room, you will turn to the right. The opposite is true if you're doing the right-handed search—you will turn to the left in that evolution. When you do come to a corner, turn the corner and crawl one or two steps forward, then move laterally toward the center of the room (fig. 7–4).

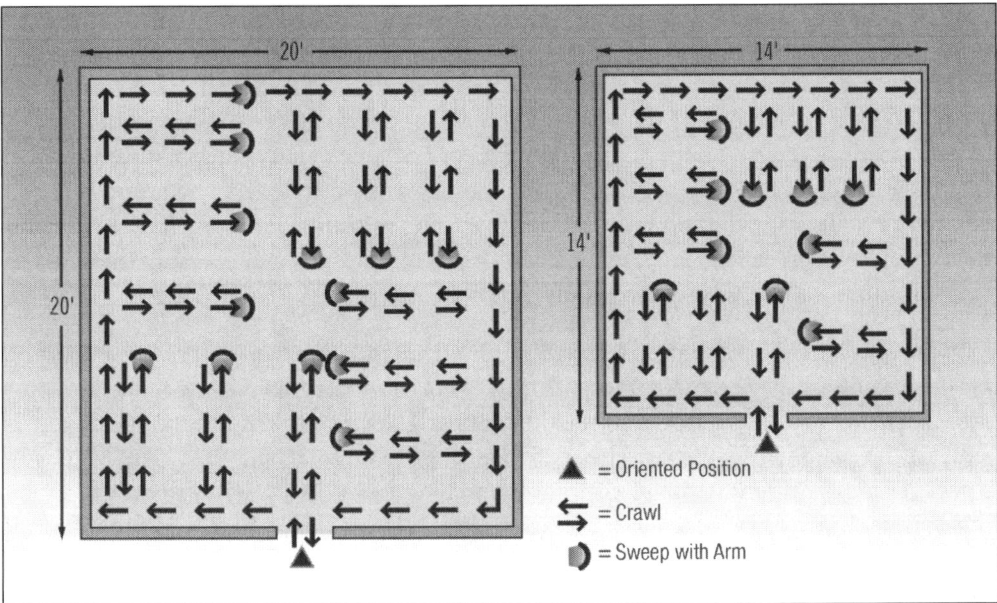

Fig. 7–4. The one-person search patterns

The number of walls in a room

This may sound a little strange, but most rooms in residential occupancies have four or five walls, depending on where the door is positioned on its wall. If the door is located in the corner of the room or at the end of a wall, then there will be four walls in that room. If the door is located more in the center of a wall, then that room will have five walls (fig. 7–5).

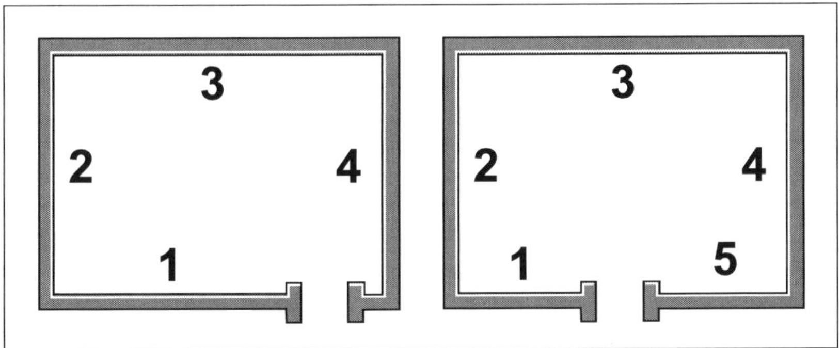

Fig. 7–5. Four- and five-walled rooms. The difference is where the door is located as it relates to the wall it sits on.

Why does this make a difference? Well, let's assume you believe every room has four walls. You are conducting a search pattern throughout the room and you have just turned the corner onto your fourth wall. You're moving along your fourth wall conducting side sweeps toward the center of the room when all of a sudden the oriented officer says, "We've just been ordered out of the building. Let's go!" I was always taught never to crawl ahead if you don't know where you're going. Logic would dictate that if you move ahead along your wall, you will come to the door, which logically will be at the end of your fourth wall. As you can see in figure 7–5, in the room on the right, the only thing in the corner is more wall. At what point would you begin to panic?

Not only is it important for the searcher to understand how many walls are in a room, it is important for the oriented officer. This person is looking out for crew safety and may have to get to a searcher quickly, so the officer needs to know the number of walls just as badly as the searcher. As the searcher enters a room to be searched, the oriented person expects searcher to report the direction of search and how many walls are in the room. This can be done easily by moving the hand along the wall on both sides of the door. If both sides are wall, there are five walls (fig. 7–6a and b). If one side is a corner, then there are four.

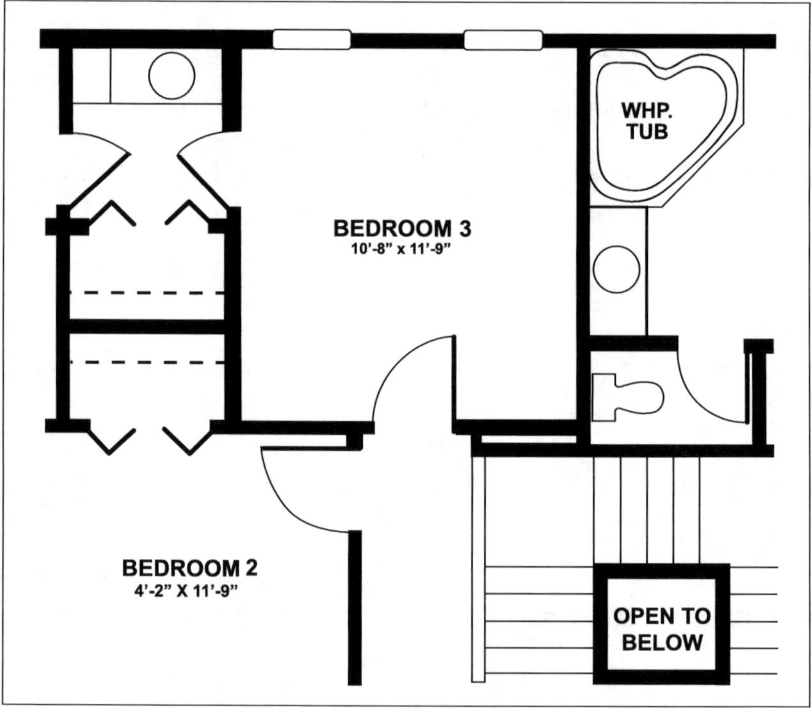

Fig. 7–6a. Determining the number of walls in a room. This is a 5-walled room.

Fig. 7–6b. Searcher enters door and checks for number of walls.

Three things required to conduct an oriented search

Proper position of the oriented officer. One of the first things required in conducting a successful oriented search is the proper positioning of the oriented officer. This is especially true if there's more than one searcher working with the officer.

The proper position for the oriented officer depends on the area being searched. In residential occupancies, when searching the second floor of a two-story home, the oriented person should be positioned in the hallway (fig. 7–7). When searching the first floor of a two-story home, the oriented officer should be positioned near the stairwell to the second floor. As you can see in figure 7–8, most areas of the first floor can be assessed and communication maintained between the oriented officer and the searcher from this construction feature. This also provides a point of reference for location in relation to the exit.

Ranch style one-story homes are more of a challenge for the oriented officer. It is difficult to find a familiar location in relation to the point of entry. A pre-connected hoseline can help maintain orientation in ranch or larger one-story homes. If a hoseline is to be used, it can be stretched dry into the structure and remain dry for ease of movement.

As a general rule in commercial occupancies, something must be used in order to remain oriented with the path of egress. Again, pre-connected hoselines work well, as do guide ropes. This will be addressed in greater detail later in the book.

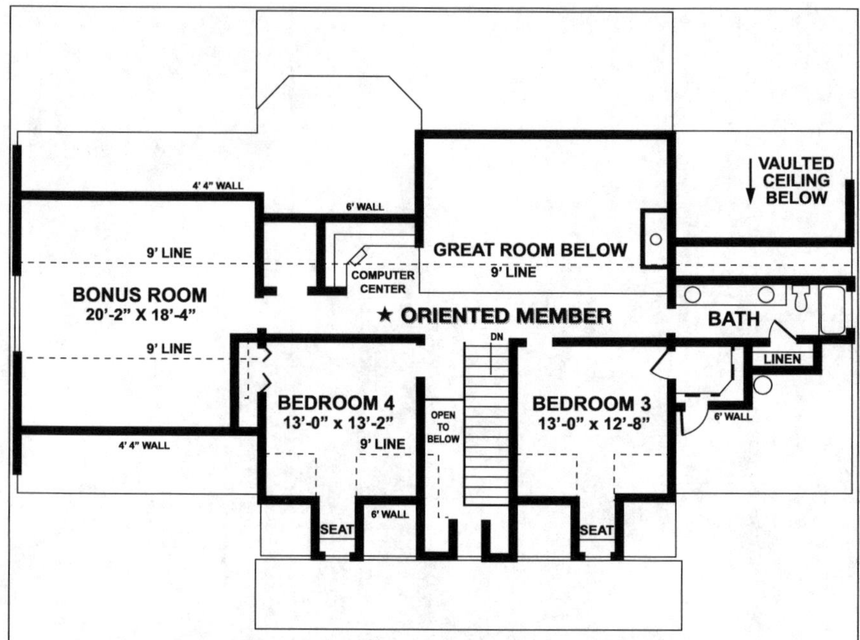

Fig. 7–7. The oriented member should be positioned in the hallway of the second floor of a house when searching the second floor.

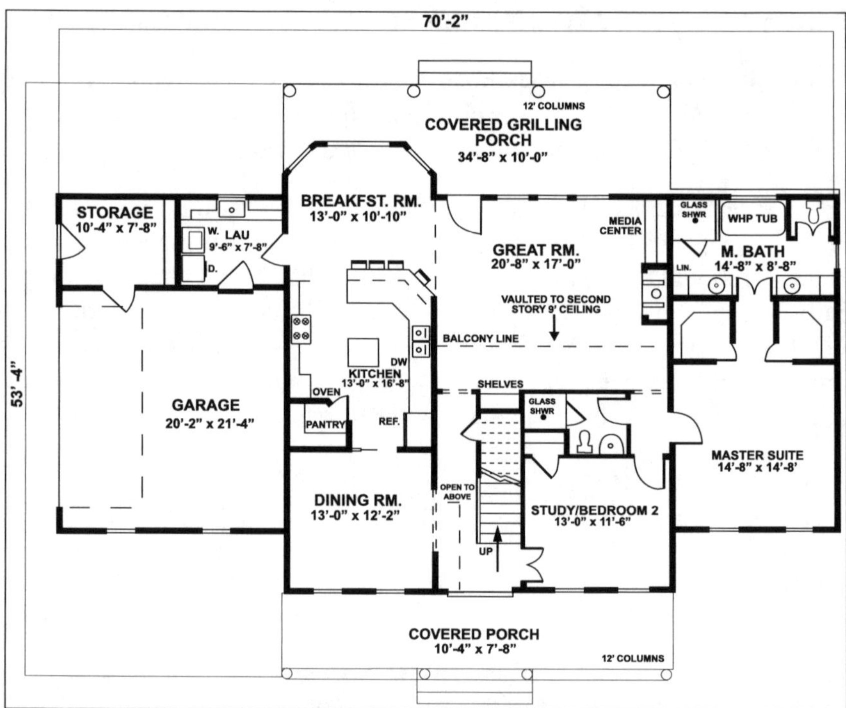

Fig. 7–8. The officer should be positioned at or near the stairwell to the second floor when searching the first floor of a two-story home.

Communication between the oriented officer and the searchers. Communication between the oriented officer and searchers is vital to a safe, effective search. Communication provides two basic elements to the search. First, it provides the oriented officer with the location of the searcher. Earlier I discussed the number of walls in a room. As the searcher moves throughout the room conducting the search, I expect the searcher to tell the oriented officer when moving to a different wall ("Lieutenant, I am on wall 3").

The searcher should also communicate anything the oriented firefighter should be aware of, such as unusual furniture in the room or where a closet or window is located. The second element provided by communication is the psychological factor. Knowing that there is someone in the hallway looking out for the searcher puts the searcher more at ease while conducting the search.

Knowing the direction of search. It is important that the oriented officer knows the direction of search. That should be communicated to the oriented officer when the searcher enters the room to begin the search. Knowing what wall the searcher is working on along with the direction of search allows the oriented member to always know the location of the searcher, even in zero visibility.

Variations to the Oriented Search

The oriented search was designed to be effective in searching both single- and multiple-family residential occupancies. However, it can be applied to commercial occupancies as well. There are a few variations required in order to make this a safe search.

- **The modified oriented search.** This form of search has as its general application searching larger apartment occupancies or large areas in ranch homes. The main difference between a traditional and a modified oriented search is that, in the modified search, as the searcher moves ahead in a larger area where communication becomes difficult, the oriented officer moves up into the structure along with the searcher. Modified oriented search usually uses teams of two firefighters: one oriented member and one searcher. If the teams cannot be split into equal numbers of two, then the oriented search is conducted, but the two searchers would use a pattern similar to that of the standard search.

- **The oriented search using a hoseline.** This method of search is generally used in smaller commercial occupancies. A 2½-in. pre-connected handline is stretched dry into the structure. The oriented firefighter is positioned literally on top of the hoseline and uses it as the point of reference to the means of egress. In essence, the firefighter has built a 2½-in.-high wall. The searchers then conduct search patterns laterally away from the hoseline in a fashion similar to that of team search. It is generally believed, when conducting an oriented search using a hoseline, that the hoseline remains stationary as the searchers and oriented officer move in their direction of search (fig. 7–9). It is not intended that the hoseline be used as an extinguishment tool, but rather as a tool to maintain orientation with the exit.

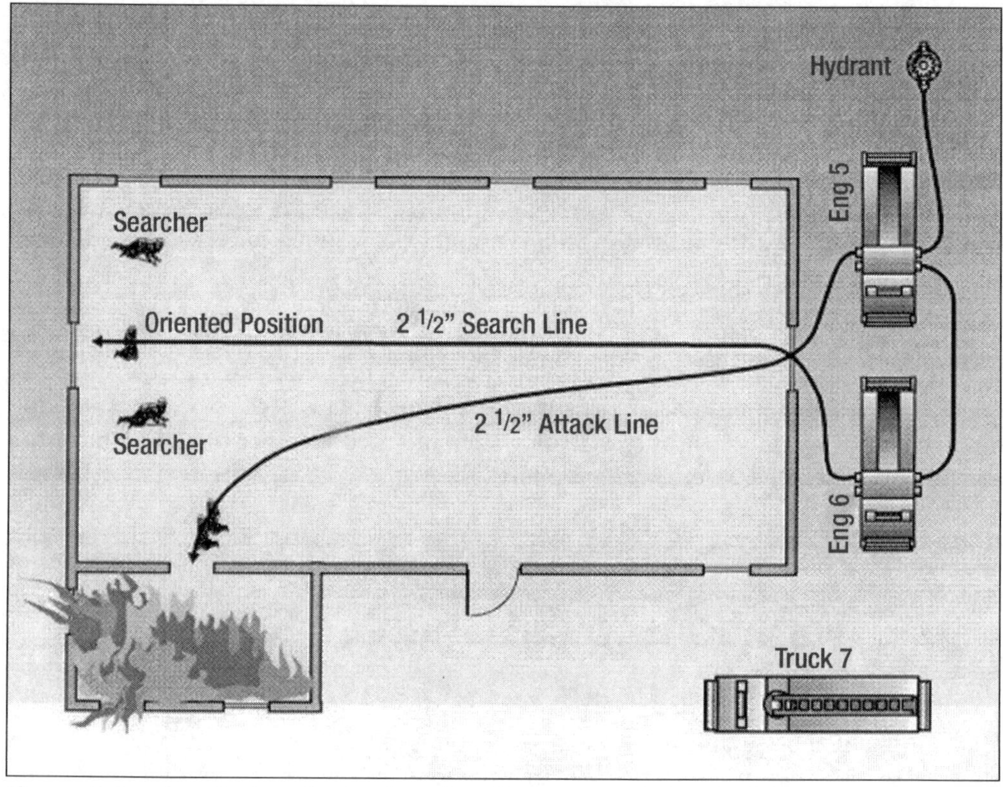

Fig. 7–9. Using a hoseline to maintain orientation with your egress in large buildings

Oriented Search and Other Groups/Divisions

The oriented member is in a great position to assist other division and group officers with their tasks. By this, I do not mean physically assisting with the attack, ventilation, or backup efforts. What I *do* mean is the oriented member can provide vital information to them as the search is being conducted and the crew moves throughout the building.

The officer conducting an oriented search and the attack officer

The oriented person should be positioned in an area where the actions of the search crew can be monitored. While this prohibits the search officer from physically entering every area being searched, the searchers can provide the officer with information concerning the fire as they search their area. I do not believe in the practice of "searching for life and fire" that some departments advocate. Each requires its own specific focus.

Searching for fire involves opening up cupboards, walls, and other areas to find hidden fire. All of this distracts and takes time from searching for victims. If, however, a searcher does find fire while conducting the search, that information should be relayed to the oriented officer who will then forward it to the attack group officer. If the location and amount of fire requires that the search in that room be stopped, this should be communicated to command as well.

As I said previously, multi-tasking inside burning buildings is generally counter-productive. I don't believe search groups should be attacking fire and vice versa.

Vent/Enter/Search (VES)

8

I was very torn about putting this chapter in the text. One reason is that I don't even know if I consider vent/enter/search (VES) an actual form of search. In its original version, it is doubtlessly the most dangerous method of searching used the fire service. Not only is it dangerous, it is against most safety standards, including OSHA standards. Departments that allow this search are literally hanging themselves out to be exposed to lawsuits and fines in the event of loss of life or serious injury to a firefighter performing these risky activities.

Having said that, as a chief officer I have assigned it to a crew in at least one instance in Toledo. At that moment, it was the difference between life and death for a civilian. It would have been interesting to ask her what her thoughts were on safety standards a few days after the fire. At the end of this chapter, I discuss a new form of VES that I wasn't made aware of. I was teaching in Estero, Florida, near Fort Myers. I learned that they had developed a form of a VES using the oriented search in combination with the original method of vent/enter/search. This new method is an extremely effective and safe way of searching the sleeping areas of residential occupancies when all indications point to savable victims in those sleeping areas.

VES Defined

Vent/enter/search was originally defined as an emergency means of searching second floor bedrooms off a porch roof for probable victims, using a roof ladder for access to the second floor from the exterior.

Vent/enter/search was developed by an East Coast fire department many years ago. The members of the department realized that more than 80% of the single-family homes in their city were two-story homes with a front porch (figs. 8–1 and 8–2). Of these homes, more than 80% had windows off the front porch that led directly into one or two of the bedrooms on the second floor. In the event of a nighttime fire with high likelihood that victims may be in one of the second floor bedrooms, they developed a procedure called vent/enter/search.

Fig. 8–1. Older house with a front porch roof leading to bedrooms

Fig. 8–2. A newer house with a front porch leading to bedrooms

Vent/enter/search is based on one person searching bedrooms alone. This in itself makes it an extremely dangerous evolution. Since its inception, its use has spread to other areas of the country and has adapted into many different evolutions.

How the Original VES Works

VES is an evolution in which one firefighter, usually the driver, sets the pumps after the engine crew has pulled and stretched the initial attack line. This by itself is very dangerous: to have an attack crew enter a building with a hoseline and not keep the driver at the pumps. The driver then removes a roof ladder from the apparatus and ladders the front porch in the event of a water of pump failure. The driver then proceeds with the following:

- Climbs the ladder to the front porch and opens or breaks (vents) the window leading to one of the bedrooms.

- Enters the room and immediately proceeds to the bedroom door to ensure that it is closed (fig. 8–3). This buys time from any advancing fire and also lessens the possibility that the attack crew will push heat and steam into that bedroom.

Fig. 8–3. A firefighter checking the door prior to beginning the search

- Searches the bedroom for any victims and leaves out of the same window that was entered.

- If a second window is accessible from the front porch, the driver would repeat this process in the adjoining bedroom. Refer back to figures 8–1 and 8–2. Both of these houses have multiple bedrooms off the porch.

That's it! As stated above, other departments have developed their procedures using the vent/enter/search name, but they allow it to be used in other applications, such as bedrooms not adjacent to attached porches.

The Hazards of the Original VES

VES is a dangerous evolution for several reasons:

- If it is conducted by the driver, then no one is watching the pumps for loss of water to the attack line while crews are inside.

- One firefighter is allowed to enter a burning structure alone.

- There are many firefighters advocating the safest fire scene possible who never recommend the use of a roof ladder as a climbing ladder. In my opinion, that "never" word should be avoided. It is true that a roof ladder has a smaller base and is not designed for climbing, but it is short, lightweight, and does work in this one-firefighter application. Still, it's dangerous.

- The firefighter is allowed to enter a burning building in the face of an advancing hoseline. This is generally not a good idea.

The Only Safe Option—The Oriented VES

There is one option for conducting VES that provides for a safer operation. This involves combining the oriented search as described in previous chapters with VES.

Staffing has become an issue across the United States. Even in departments where staffing was low to very low, buildings have gotten bigger and homes have become more complicated, larger, and constructed of cheaper materials. Many departments now conduct searches using two firefighters, and many of those departments pull up to working house fires with 10 or fewer firefighters. In those extreme and rare instances where a rescue is imminent, especially at fires in the night time hours, VES provides a rapid method of searching bedrooms in residential occupancies.

In order to safely conduct VES operations I recommend the following.

Two-story home

A two-person crew proceeds to the bedroom area from the exterior of the building. A roof ladder or a ladder less than 28 ft. can be used on two-story homes to reach the second floor or division two. If the home is an older traditional home with a front porch that is used to gain entry into the sleeping areas, in these dire times, a roof ladder can be used to access the front porch roof. You must remember that if a victim is located, the victim can be brought out to the roof, and a ladder of significant size, a tower ladder, or a bucket can be raised to remove the victim.

A window to the bedroom should be opened or forced. A firefighter then enters the bedroom to locate the interior bedroom door and close it. Then the entire bedroom area is searched. While this is done, the other firefighter (the oriented officer) remains outside on the front porch performing the duties of the oriented position. In this way, no single firefighter is ever performing search entirely alone.

If the only access to bedrooms in is in the rear or on the side of the home, then, if time and staffing permits, a climbing ladder should be used to gain access to the division two bedrooms from the exterior. The searcher would proceed first up the ladder, open or remove the window, and then enter as described above. The oriented member would climb the ladder and be positioned on the ladder at the window, but remain on the outside (fig. 8–4).

Fig. 8–4. The oriented member generally remains at the window outside while the searcher searches the bedroom.

This process should continue until all the sleeping areas have been searched. Ideally the fire will have been knocked down by then. Conditions will improve to the point where the remainder of the house can be searched from the inside using safer, more acceptable methods.

One-story home

The two-person crew would proceed to the sleeping areas. Attic ladders could be used on one-story homes. There are departments that would balk at using an attic ladder on a one-story home, saying that it slows down search efforts and that the second firefighter can easily boost the searcher through the open window (fig. 8–5). This action generally requires that the searcher rapidly maneuver over the open window and jump onto the floor below. I do not advocate jumping on any floor in a building that's involved in fire. This is especially true if the fire volume and conditions indicate that VES is a necessary option. With today's lightweight construction becoming more and more prevalent, we have to avoid putting major stress on any structural members at a well-involved fire. Other means of entry include boosting the firefighter into the window and using a Halligan bar or similar tool with Prusik knot to assist in gaining entry (fig. 8–6).

Fig. 8–5. A two-member crew starting an "oriented" VES

Fig. 8–6. Using a Prusik knot on a roof hook to gain entry to the first floor bedroom

The attic ladder should be placed at the window. The first searcher climbs the ladder and opens or removes the window, then

- enters the bedroom and immediately locates and closes the interior bedroom door;

- conducts a search of the bedroom and exits the window using the attic ladder.

While this is occurring, the oriented member remains at the attic ladder and stays positioned outside the building in the event that a victim is located who needs assistance and removal, or if the searcher gets into trouble. This process is repeated until all of the sleeping areas in the house have been searched. At that time, more traditional searches could be considered.

Under Extreme Conditions Only!

One-member VES operations violate most accepted safe practices in the fire service. Two-member VES operations provide a safer operation for several reasons:

- Although one firefighter has entered a burning building alone, there is another firefighter within voice contact, acting as the partner should the searcher gets in trouble. I could argue that this maintains the buddy system and does not violate the concept of "two in, two out," or the rule that one firefighter should not enter a building alone.

- As one firefighter concentrates on searching the bedroom (and only the individual bedroom with the door to the rest of the home closed), the oriented member can maintain voice contact with and account for safety conditions for the searcher. The oriented member can monitor heat conditions above floor level by exposing his or her wrist through the window.

- Under extreme fire conditions in either a one- or two-story home, the oriented member could enter through the window into the bedroom, but remain at the window to better assess safety conditions within the structure. The stability of the floor as well as the amount of heat at floor level and on the floor surface could be monitored. The warning here is that, in the event immediate withdrawal is required, two firefighters must bail from the bedroom window. If these rare extreme circumstances occur, the oriented member orders the searcher to leave the structure, then immediately exits the window to descend the ladder so as not to delay the egress of the searcher.

Extreme caution must be used if you believe that the room you intend to search using VES is also the fire room. Under most circumstances, the first evolution performed is to put out the fire. This necessitates that a hoseline is stretched into the building and to the area of origin. If this area of origin is a sleeping area or a bedroom and the door is closed, common sense would dictate that the attack crew will open the door to affect extinguishment. Another firefighter in that room performing VES could be in serious jeopardy from the extinguishment process. If viable victims are in the fire room, the attack crew should locate them after knockdown. Under most circumstances, I would not use VES in the room of origin or where the fire is upon arrival. It is simply too dangerous for the searcher, whether alone or using the oriented method of VES.

One last time, I must emphasize that this is an *extremely* dangerous tactic. I have always believed that drastic times call for drastic measures. If I believed a save could be made using this method, especially the life of the child, and I had a firefighter with the knowledge, skills, and ability to perform this evolution, I wouldn't hesitate to direct that member to do so. I believe that if it was your child up there, you would want a firefighter to try this. However, I am aware of too many departments that take some form of this evolution and use it at everyday fires. This is an evolution that in my opinion may be used once in a career to save a life and should not be practiced or taught as an everyday evolution. Two firefighters, one acting as a searcher and the other as the oriented member, should always be used if this tactic is to be considered.

Primary and Secondary Searches

9

I wish the people who wrote fire service textbooks could get their act together. That includes me, as well. The problem I see all the time is that the fire service tends to be very parochial. We all think we do things differently in "our department" than other departments do across the United States. On the surface, that's not a bad thing. What works in Toledo works in Toledo. What works in Tulsa works in Tulsa. Problems start to arise when we put our thoughts and ideas on paper for others to read.

You read one textbook that calls a specific evolution one thing, then read another and it is called something different. One author will define it one way and another has an entirely different definition. It drives me crazy when I come across two different textbooks with different definitions for the exact same word.

I say all this because a lot of what I discuss in this section of the book may differ in wording from how other textbooks explain it. I'm not going to say I'm right and they're wrong. Just use what works best for you. If you don't like the name, phrase, or definition of something that I'm talking about but you *do* like the evolution, then call it whatever you want. The only thing that really matters is that you, your crew, and the crews that respond to fires with you understand what your terms mean.

Primary and Secondary Searches

I'm no historian, so I can't tell you the origin of primary and secondary searches. When I was a fire recruit back in 1975, I wasn't taught the concept of secondary search. It just didn't exist. I believe the first time I came across the term was in 1978 when I was studying for my first promotional exam. My understanding of primary and secondary searches didn't always mesh with what I was reading in textbooks. Here are the definitions I've come up with after years of conducting and supervising searches on firegrounds of all sizes.

Primary search

I define primary search as a rapid, systematic search to locate savable victims. The operative word in that definition is "savable." Because of the by-products of combustion, including heat and smoke, we only have a limited amount of time to locate and remove victims from IDLH atmospheres. We can make estimations about the number of potential victims inside a structure based on occupancy type and information from witnesses. However, even if the neighbors say there is only one person living in that house, we should always assume that there may be others inside, as well.

When conducting a primary search, if we locate an obviously deceased person, I believe that information should be transmitted to the incident commander along with the location of the victim. After that, the search team should continue with the original search pattern looking for viable victims. They need to finish the original search until the entire IDLH area has been covered.

I speak from a place of experience when I tell you that this is a hard thing to do. Our natural instinct when we locate a body inside a burning structure is to give that individual the benefit of the doubt by removing the victim and beginning resuscitation efforts. However, this is not always the correct thing to do. One of the angriest human beings I've ever seen was a coroner when one of my crews removed an obviously dead individual from a house. We later learned that the deceased was a victim of a homicide and we'd made the investigation that much harder.

I always tell my crews that if there's any doubt as to the viability of a victim, then go ahead and bring the victim out and we will start resuscitation efforts. However, if the victim has obviously passed on, then continue your search for savable victims. You would hate to leave another live victim inside a building while you're busy carrying out a corpse.

Primary searches should be conducted with a plan in mind. They should systematically cover all areas that are IDLH in the involved area or structure. Areas that are not IDLH can be covered later. With primary search, time is the motivating factor. Seconds can mean the difference between life and death.

Secondary search

When I first encountered the term "secondary search," it simply referred to searching the same area a second time. The only difference was the use of different crew members for the second search.

My definition of secondary search differs quite a bit from that original definition. I define it as a slow, methodical search to locate deceased victims.

My definition is built on the premise that the primary search was conducted correctly the first time. It assumes that no time was wasted duplicating efforts in previously searched areas while other areas are still waiting to be covered. The clock is always ticking, so do it right the first time!

Secondary searches are conducted after *every* IDLH area has been initially searched. They should take place with much better visibility than primary searches. Hopefully the area being searched has been well-ventilated and lights have been set up. This reduces the chance of injury to firefighters operating in the area. Depending on your department's policy, secondary searches may even be done without the use of SCBA, as long as proper ventilation has taken place. They should always be conducted in a well-planned, logical manner that keeps the safety of the crews conducting search as the highest priority.

Fig. 9–1. The bedroom on the B/C corner of this Victorian house has reached flashover. (Photo by Lt. Bob Pressler, FDNY [ret.])

Primary searches are rapid systematic searches to locate savable victims. Secondary searches are slow, methodical searches to locate deceased victims. In essence, the difference between the two is the condition in which you expect to find the victim. With primary searches, your time and effort is used to locate savable victims. There are several factors that apply here. Among them are the location and the intensity of the fire, as well as its proximity to savable victims. In figure 9–1, the bedroom on division two in the B/C quadrant has reached the state of flashover. I would not expect a crew to conduct a primary search in this room. If we can only go by observable conditions from the picture, then the remainder of the second floor would call for a primary search. Again, considering only the information gained from the photo, a secondary search could be conducted in one of two ways for the room involved in flashover. One way is for the attack crew to conduct a quick search immediately upon completion of knockdown *if* they believe conditions would allow a savable victim to still be located in the fire room. If the attack crew, after knocking down the fire in the room involved in flashover, believed that fire has gotten into the attic or the area above, they likely would take their line to that adjacent area to continue extinguishment. Under this circumstance, I would expect a crew to do a secondary search in the room that was involved in flashover after ventilation had taken place and visibility improved dramatically.

Prioritizing Search

10

Staffing is becoming more and more of a concern to fire departments across the United States. As of this book's writing, the economy is going through the second year of a recession that has brought layoffs and cutbacks to departments that never envisioned such an event. Most fire departments do not get what they consider an adequate number of firefighters arriving initially to the scene of a fire. Currently, the standard (NFPA 1710) considers 13 firefighters the minimum adequate staffing for a residential structure fire. To some metropolitan departments, that number seems painfully anemic. To others, however, having 13 firefighters at the scene of a house fire would be a godsend. NFPA 1720, the companion standard addressing volunteer and combination departments, calls for a minimum of 10 firefighters to arrive at a fire in 10 minutes. Again, for most, this is a daunting if not impossible task. Many departments get 10 or fewer firefighters to respond to a reported fire in a residential occupancy, or any occupancy, for that matter.

In my previous department, when I was on the job, we averaged about 28 firefighters at a report of a house fire. In the inner city area, where fire stations were built in relation to how far a horse could run, most of those firefighters got there at about the same time. In the outlying parts of the city, we traditionally would get a unit to arrive, then another unit a minute or two later, and so on until the full complement of five engines, a truck, a heavy squad, an ALS life squad, a safety officer, and a battalion chief would arrive. Because not everyone arrived the same time, we were forced to do sequential fire operations.

Sequential fire operations are based on the premise that only a certain number of firefighters are available to conduct the myriad of tasks required at a working house fire. For that reason, operations had to be performed in a sequence based on priority until adequate staffing allowed for simultaneous operations to be conducted.

Prioritizing Search

It is the responsibility of the incident commander to determine when searches are to be conducted. I believe this holds true for every operation on the fireground. The "hows" and "wheres" of the search are generally left to the officer assigned that task. When to prioritize search is, at times, an extremely hard decision. The decision should be based on several factors.

"

- **The probability of occupants being inside.** It would be my guess that, of the number of firefighters we lose each year, several of them died looking for victims who were never really in the building. We should act on the information received from occupants and their neighbors concerning the probability of victims still being inside. We should also read the visible indicators as we arrive at the scene of a fire. Those include cars in driveways and garages, toys and other belongings on the front porch and yard, as well as the general condition of the building. The likelihood of finding a savable victim in the building in figure 10–1 would be remote. In this instance, even if a bystander "thought" there was someone inside, I would not conduct a primary search absent "seen or heard" victims until the fire was darkened and the building vented. I would still direct crews to conduct a search, but I would do it on our terms. Wait until there's good visibility for the search crew after ventilation efforts have cleared most, if not all of the building.

- **Size of fire.** A building totally involved in fire or flashover reduces the likelihood of a civilian (or firefighter) surviving the fire (fig. 10–2). The rule of thumb is the more fire involvement, the less likely it is you will find savable victims. If only one or two rooms are heavily involved, it increases the chance of survivability for victims in areas adjacent to or remote from the fire rooms. There's not a hard-and-fast rule here. Experience assists the decision-making process.

- **Location of fire.** My rule here is "The lower the fire is in relationship to the building, the greater the hazard to occupants." Heat and hot gases rise. Basement fires in occupied residential buildings present the greatest potential for trapping civilians. Attic fires present the least potential.

Fig. 10–1. An obviously vacant house

Fig. 10–2. Savable victims are not likely to be found in a structure this fully involved.

- **Location of fire as it relates to occupants.** The location of fire as it relates to occupants can present challenges when prioritizing search. Attic fires generally present the least potential of injury to occupants. If possible victims are in a room adjacent to the fire area, search should be moved up the incident commander's list of priorities.

- **Training of the crews available.** A good chief and company officer will know their people. They know their limitations and their abilities because they drill with them on a regular basis. Any person tasked with sending firefighters into a burning building must be aware of the knowledge, skills, and abilities of their crews. How else can you make sound judgments on what they will be doing inside? This is a two-pronged approach. Participating in search drills is part of the approach. The second prong is conducting tailboard critiques immediately after the incident is under control. After an incident, the IC should always ask the officers and crews, "What did you see? What did you do? What was your thought process?" By doing so, they will gain a solid understanding of what is happening with crews inside the building. That's the only way to learn what they are really doing inside while the incident commander is standing outside. At those very rare fires that actually involve life and death circumstances, send your best, most seasoned crews in on search. We are called upon to become part of the solution to a problem. We let both the citizens and ourselves down when we become part of the problem instead.

- **Staffing on-scene.** The last factor in prioritizing search deals was staffing. This is why each chief should know what the crew's abilities are. This holds true for company officers as well. They need to know what the crew and the officer can realistically accomplish. This holds true for residential occupancies as well as commercials. When contemplating search in commercial occupancies, as you will see later in this book, the incident commander and search officer need to make realistic assumptions of what their crews can accomplish. We can't search forever in IDLH conditions. Can a single four-person crew, including the officer, search a restaurant in zero visibility IDLH conditions and still hope to pull out viable victims from the last remaining area to be searched? If you pull up to a working fire with a probability of victims being trapped inside and only have enough personnel on hand to successfully accomplish one task: put the fire out! If in doing so you stumble upon a victim inside and staffing is still not available, you have an extremely difficult decision to make. That decision becomes even tougher if you believe there is more than one victim inside. This goes back to the rule "do the best for the most." Said one final but different way: if staffing is low and the choice is between putting out the fire and searching . . . put the fire out!

Prioritizing Assignments

There are several ways to prioritize assignments at a fire. Many chapters in fire service textbooks have been dedicated to the subject. One such method is what I call "Chief Brunacini's Way."

Chief Brunacini's way

Chief Alan Brunacini was chief of the Phoenix fire department for more than 40 years. He has taught and lectured all over the world and written several fire service text books. In *Fire Command* (1985), he discussed the four priorities at every incident. Those are:

1. **Firefighter safety.** Our first priority is the safety of our crews. We are sent to be part the solution, not part of the problem. If we neglect safety concerns and one or more of us becomes injured or threatened by the fire, then we become part of the problem. When I pull up to any incident the first question I always ask myself is, "Can we safely and effectively operate in the current environment?" If the answer is "yes" then I move on to the next priority. If the answer is "no" then I believe we either have to change fire conditions or wait until conditions change themselves.

2. **Civilian safety.** Once it is determined that our firefighters can safely and effectively enter the structure to commence firefighting efforts, our next priority becomes the safety of civilians. We should do what is necessary to protect any citizen endangered by the fire. This can be done in several ways. With a lack of seen or heard victims, we accomplish that goal best by putting out the fire as rapidly

as possible. In the presence of seen or heard victims, a priority decision needs to be made. There is a line between victims who are actually endangered and those who don't face imminent danger to their life and health. A victim observed from a window in the fire building, remote or even below the fire, with no or little fire issuing from the window, will need to be calmed and assured, but unless their peril worsens rapidly and dramatically, they can be left until additional staffing arrives. The officer will have to make a decision in those situations depending on where they feel they can be most effective. Will pulling the victim out cause the greatest good for the most people, or will stopping the fire spread to other areas help the most victims? It can be a tough decision to make.

3. **Stop the problem.** Our job is to be part of the solution. That's why people call us. If it's a fire, let's put the fire out. If it's an EMS run, let's triage, treat, and then ensure transport by whatever means available.

4. **Conserve property.** According to Chief Brunacini, our last priority is to do the three above while conserving as much property as possible. Life always comes first, but do what you can to save victims' belongings.

R.E.C.E.O.

Another method of prioritizing assignments at incidents uses the acronym R.E.C.E.O. First off, let me say that I do not like using acronyms in any way, shape, or form. Try remembering what the second "E" in R.E.C.E.O. stands for when you haven't thought about it for several years and have people hanging from windows at a working fire. However, it is a commonly used method so I have included it in the text.

The acronym stands for:

- **Rescue.** The first letter in the acronym stands for rescue. This indicates that our first priority should be attempting to rescue threatened civilians at the incident.

- **Exposures.** The next priority under this system is to protect exposures by confining the fire to the smallest area possible. This is closely related to hoseline and nozzle placement.

- **Confine.** Begin to control the BTU production to a point where you are cooling the fire more than it is producing heat. Some texts refer to this as "containment."

- **Extinguish.** Darken down the fire.

- **Overhaul.** Ensure that the fire is completely out.

One of my concerns with this acronym is in teaching it to young, inexperienced firefighters. When they are told that our first priority at every fire is to rescue endangered occupants, they immediately think of firefighters running into a burning building with the sole intent of locating any possible victims. That is generally not an accepted practice at the vast majority of fires to which we respond.

Common sense approach to prioritizing

The last method of prioritizing your assignments at a fire is what I call the common sense method. This method goes back to the chief or incident commander knowing the capabilities of the firefighting force. It entails:

- **Staffing constraints.** What is the actual number of firefighters that you have to work with? In some major metropolitan fire departments that number is about 35 firefighters who all arrive around the same time. In most metropolitan fire departments, I would guess the number is somewhere between 20 and 30 firefighters who will arrive one unit at a time. In rural America, the number is anywhere from 2 to 20. Most fire departments use sequential firefighting operations. This means that staffing prohibits multiple or concurrent tasks from taking place. Crews are generally tasked with conducting an initial assignment and then moving on to another assignment. This is all orchestrated by the incident commander.

- **What crews can actually accomplish.** Can your three-person engine crew bring an initial fire flow requirement of 750 gallons per minute (gpm) in the interior of a warehouse? Can your truck company complete a trench cut on the roof of an apartment building in time to keep a fire from spreading to an adjoining section? Can a single four-person crew search a 12-unit section of a garden apartment building in time to pull out any and all victims? It is incumbent upon chief officers and incident commanders not only to know what they can perform, but the time frame and their intended output (as in gpm).

- **Text book practices.** Do your crews know the evolutions you intend them to complete? This is a training function! One of my major complaints with training in the fire service is that recruits are not required to read the textbooks that the older firefighters used to study for promotional exams and reference books and vice versa. I can say with all honesty that I have never read, from cover to cover, the recruit manual used by my previous department. I read a few excerpts, and disagreed with a lot of what it said. However, its use is mandated by the state.

- **Witness stand concerns.** The United States is becoming a very litigious society. It is not uncommon for fire departments and municipalities to be sued for negligence. Although not in the forefront of my mind when running a fire, I do occasionally ask myself, "Can I defend what I'm doing on the witness stand?" Is an on-scene force of 10 firefighters with two engines and an aerial enough to put out a well-involved fire in a 100,000-sq.-ft. warehouse, or should I request additional crews?

Golden nuggets

I would like to provide you with a few "nuggets" related to prioritizing assignments at a fire:

- It's not a perfect science. You will make mistakes. Just move past them and worry about the issues in front of you.

- There's always more than one "right way" to fight a fire. Conversely, there's always more than one wrong thing to do at any time during a working fire!

- Always ask yourself, "Is it as safe as possible? Is what we're doing worth the risk?"

- Try to ask yourself, "Could I sit on a witness stand and defend our actions?" In today's litigious society, you need to be able to explain your decisions on the fireground.

How I prioritize search

Let's look at some illustrations to give you an idea of how I prioritize search.

Search is definitely a high priority at the fire in figure 10–3. The fire appears to be concentrated on division two in the B/C corner bedroom. This appears to be a daytime incident with the fire being concentrated on division two. My first priority would be to get an attack line positioned at the doorway to that room involved in fire. Assuming adequate staffing, I would then simultaneously start search and ventilation efforts.

Fig. 10–3. Partially involved house fire. (Photo by Lt. Bob Pressler, FDNY [ret.])

In figure 10–4, search would be a very low priority to me. Considering the occupancy type, the amount of fire, and the time of day, I would not expect to find viable victims in any areas where search would be possible because of fire conditions. I would also logically expect that any civilians in exposed occupancies in the strip mall would have self-evacuated by this time.

Search would not be a priority to me at the fully involved house in figure 10–5. No human being, including firefighters in their bunker gear, would survive more than a few seconds in that environment. Conducting secondary search efforts would be questionable due to the amount of fire and the stability of the structure after knockdown.

Fig. 10–4. Heavily involved strip mall. (Photo by Tod Parker, http://www.PhotoTac.com)

Fig. 10–5. Fully involved house fire. (Photo by Lt. Bob Pressler, FDNY [Ret.])

11

Search vs. Rescue

I've been associated with fire service for more than 35 years. During that time I've read many textbooks and magazine articles and watched hundreds of videotapes and news accounts concerning the fire department and fire department operations. There are many books written that use the terms "search" and "rescue" together as if they were one operation. I often hear accounts from firefighters discussing fires where they conducted "search and rescue" operations as if they were one and the same. I believe that they are two separate and distinct operations. It all goes back to focus.

Search Defined

Search is defined as "a rapid, systematic evolution designed to *locate* savable victims in a structure involved in fire." There are several different methods that can be used to obtain this outcome. Once again, it goes back to focus. The focus of the search crew must be to move logically about an area filled with fire and heavy smoke with the objective of locating endangered occupants.

Rescue Defined

Rescue is defined as the process of moving located victims from an area that is IDLH to an area that is not. This area that is not IDLH can be outside of the building or in a safe haven inside the building involved in fire, depending on the circumstances. This would be acceptable in large apartment buildings and high-rise structures.

Few departments create plans for specific evolutions for rescue. There are several reasons for this. One is that we actually rescue civilians at very few of the fires to which we respond. I was on the job 32 years. During those 32 years, I only spent 12 riding an apparatus as a firefighter, lieutenant, or captain. In those 12 years, I had two actual rescues and one "sort of" rescue.

In one of those rescues as a captain, I entered the house with a fire in division 2, without SCBA, while my crew pulled a 1¾-in. handline behind me. Civilians on the scene indicated a likely victim on the second floor hallway just past the room involved in fire. I ran upstairs past the fire room, grabbed the victim, and brought him down the stairway as my crew was bringing a line up.

In another rescue, civilians believed a child was in the first-floor bedroom at the rear of the house. There was visible fire on the AB corner and heavy smoke throughout the house. I told my crew to pull a line and get it on the fire. My intent was to mask up and join my crew, but civilians grabbed me and took me to the back of the house to the bedroom where the child was believed to be. They had opened the window in an attempt to remove her but were unsuccessful. I had them boost me into the window. I entered the room an immediately found the child on her bed and brought her out the window.

In my "sort of" rescue, I located a family in an apartment building involved in fire. The room they were in was two floors above the fire floor with light smoke conditions in the hallway. There were two adults and an older child in the apartment. I took them to an adjacent stairway and out of the building.

Another reason rescues are hard to train for is the myriad of factors that could be involved. Building configuration, occupancy type, and the potential number of victims are all hard to anticipate. A lot of firefighters are trained as recruits in basic victim drags and carries, but generally that is the extent of our rescue training.

The Problem

When you are assigned search, your focus is on *locating* lost, trapped, or endangered civilians. Once located, you redirect your focus to the *rescue* of that civilian. They are different evolutions, utilizing different tactics, and have different results.

Once you locate and begin to *rescue* the victim, where is your search? It's gone. At that split second, all of your thoughts about conducting an organized, logical search vanish and you are immediately consumed with the process of moving the victim to a safe location. The problem is that your search for additional victims ended at that time. That is when we run into issues with continuity of search.

Another problem that arises is simply the math. If firefighters are allowed to remove victims and remain with them, then the number of firefighters required to successfully remove all potential victims becomes extremely large. For example, imagine a working fire in a single wing of a one-story nursing home. Let's assume that the wing involved has 12 rooms dedicated to patient care. Let's also assume that each room is designed for two elderly patients. That gives us a total of 24 potential victims in the entire wing. If two firefighters remove each victim and remain with them instead of passing off the victim and returning to their search, then 48 firefighters would be required to locate and remove, then care for, all the victims. How many firefighters do you get initially at the report of a fire in a nursing home?

The Solution

The issue of separating search from rescue is not an insurmountable problem most of the time. The vast majority of us respond to single-family residential occupancies. In these types of incidents, the number of victims and the area to be searched is usually within an acceptable proportion for a crew to handle. Discipline is required to maintain continuity of search.

Continuity of search is defined as assuring that the search has been conducted in a logical and uninterrupted manner. It entails knowing where we have searched, where we are searching now, where we will search next, and when the entire area has been covered.

This becomes much more difficult when we find and remove victims. It is a rare occurrence when a firefighter or crew locates and moves a victim, then returns to the exact area where the victim was located to continue the search. On the contrary, what generally happens is the firefighter or crew locates and moves the victim, and then begins lifesaving efforts. Even if advanced life support crews are outside to receive the victim, that crew will rarely re-enter the building. I've been told there is a psychological reason for this. A bond between the rescuer and the victim develops and it is difficult for the rescuer to leave the victim until he or she is safely transported to a hospital.

The responsibility for eliminating the tie between search and rescue falls in two places. First, it is the responsibility of the fire department administration and its training bureau to ensure that firefighters understand the difference between the two evolutions. Second, it is the incident commander's responsibility to ensure that in instances where victims are probable, staffing is available for the establishment of a rescue group. They should also make sure that searchers return to complete the search after a victim is removed from the building.

Rescue Groups

The rescue group is defined as two firefighters in full protective gear staged outside the IDLH area with the intent of removing located victims to the outside to receive medical treatment. Upon passing the victim off to the medical personnel, they should return to their staging area and wait for another victim to be located.

How it works

Upon locating a victim, a searcher notifies the search officer and begins to remove the victim from the area. The search officer notifies the rescue group via radio and tells them where the victim is being taken, such as the stairway or the rear door. The rescue group moves to that area, dons their SCBA, and begins to move inside to meet the searcher with the victim. When they meet, the rescue group brings the victim back outside and the searcher returns to the area where the victim was located and continues the search.

When to establish rescue groups

A well-disciplined search officer and crew rarely need the establishment of a rescue group at a single family house fire. The operative phrase is well-disciplined. At the rare fire where victims are located, generally only one or two victims are found (fig. 11–1). The officer must not allow the firefighter who located the victim to remain outside after the victim is removed from the building. The officer must ensure that the firefighter returns to the area where the victim was located. Only then can continuity of search be guaranteed.

- **Apartments and other multi-family occupancies.** A rescue group should be established at all fires at apartments or multi-family occupancies. This is especially true of four-family occupancies or larger, garden apartments, and other apartment buildings. In the case of duplexes, triplexes and four families, it is at the discretion of the IC.

- **High rise (residential & commercial).** A rescue group should be established at any working fire involving a high-rise building. The rescue group should be staged a floor or two below the fire floor in a non-IDLH area. Due to the size of the building and potential number of victims, more than one rescue group may need to be established. I also suggest that initial emergency medical service (EMS) or a lifesaving treatment area be set up as well.

- **Hotels and motels.** A rescue group should be established in much the same way as a high-rise building, whether the structure is high-rise or low-rise. The potential for large loss of life in a working fire in these occupancies is high. Rescue groups should stage at the most appropriate area of egress, closest to the area where the search is being conducted.

- **Night clubs and restaurants.** A rescue group should be established at any working fire in a nightclub or restaurant where it is believed that several patrons are present. A rough estimate of the number of potential victims can be made using the number of vehicles in the parking lot, along with information gained from civilians in the area. Some may have escaped before your arrival. Any information they can give should receive the highest priority.

Fig. 11–1. At the Norwich fire, two of the seven children are seen being carried out by search crews

Who can fill the role of a rescue group?

Any member of a rescue group must be trained in the use of SCBA. They will be called upon to enter areas that are IDLH, even if only for short periods of time. It is also useful for members of a rescue team to have EMT training. They can begin initial triage efforts on victims before they've been completely removed to the staging area. I realize that keeping a rescue group available requires more personnel. However, failing to establish a rescue group may mean the lives of civilians at those rare but deadly fires.

Reading the Building for Search

12

One of the biggest problems we have today regarding search is that the officer fails to create a search plan. That plan should include where to start, where to stop, how are we going to enter, and what type of search will we conduct once we are inside. This can't be pre-planned because the factors that go into your decisions change from fire to fire and building to building.

It is imperative that an officer creates a plan upon arrival at a fire when assigned search. Part of the plan can be constructed or developed by using a process that I call "reading the building for search." These are easy skills to acquire and every firefighter should practice the art of reading the building for search. Doing so correctly can save precious seconds or minutes in reaching savable victims.

There are six components to reading the building for search. Of the six, four can be practiced on every emergency medical run, or any other run for that matter. I know there are those out there who complain that fires are down. First of all, we should thank God that fires are down! Fires kill 1,500 children every year in the United States. I realize the fact that the lowered number of fires is a double-edged sword. Although we're aware that a reduction in fires means potential savings in lives, we also aren't as honed in all our skills as firefighters were in the past. So, as you respond to your next EMS call and you and the crew are walking up to the house of Mrs. Smith for the third time this month, turn around and ask the recruit, "Where are the stairs?"

Use every opening to train for search, even if you're on a routine non-emergency call. Since we get fewer chances to practice our search skills in live situations, we have to use every opportunity we have.

The Six Factors of Reading the Building For Search

Construction type

This text doesn't concentrate on building construction. However, it is imperative that firefighters have a basic understanding of how buildings are constructed. I believe I would be negligent if I didn't provide some basic construction information as it relates to search.

There are several ways that firefighters classify buildings. These specific classifications have been developed in order to assist firefighters in ensuring that buildings are built as fire safe as possible. For our purposes, I believe the easiest way to gain an understanding of the building's construction is to define construction by the building material used for load bearing walls.

There are four standard materials that are used to construct load bearing walls. When I say "construct buildings," I am talking about the material used to transfer the weight of the structure to the foundation of the building by way of load bearing walls. If we can determine what the load bearing walls are predominantly constructed of, then we can make educated decisions on how that building will react in a fire. Please remember this is Building Construction 101: Down and Dirty Basics.

Concrete construction. Concrete construction can be made to deliver any amount of fire resistance required by the architect. In other words, concrete construction can be made to be totally fireproof. There are two basic types of concrete construction: "cast-in-place" and "precast." It's really pretty simple to differentiate between the two. Cast-in-place concrete is poured and allowed to cure or set in the place it is intended to remain. Precast concrete is poured and allowed to set in one location, then transported to the area where it will become part of a structure. The foundation of a new home built today will generally be a cast-in-place concrete wall. Many new strip malls and commercial occupancies built today are precast assemblies. Tilt up construction is one common method of precast construction.

The good news about concrete construction is that it can be made to be virtually fire resistant. The bad news is that when a fire is allowed to develop in a concrete building, the structure itself generally retains a lot of heat. This type of construction can lead to some very hot fires.

When discussing precast concrete construction like tilt up assemblies or parking garages, we must pay particular attention to the connections that are holding the precast concrete assemblies together. These are generally made of steel. Steel begins to fail at 800°F after about 5 minutes of exposure. That is not a very hot fire.

Steel is there because it has to be there to hold the precast concrete pieces together. If attacked by heat, these steel connectors can fail. That may lead to failure of a portion of, or the entire assembly.

Steel construction. When we think of steel construction, two types of buildings come to mind. The first is smaller steel buildings like large garages, auto repair shops, and manufacturing businesses. They are built with the exterior load a bearing walls constructed of steel columns. Corrugated metal wall assemblies are then riveted to the steel columns. These buildings are very susceptible to early collapse.

The second type of steel construction is larger buildings such as high-rise or large, low-rise buildings. Their exterior load bearing walls are supported by steel columns. The final weather proofing layer applied to the steel columns can be brick, glass, stucco, or a wide variety of other exterior wall surfaces. With proper fireproofing of the structural steel columns, roof, and floor assemblies, these buildings can also be made highly fire resistant. Other than the deadly attack on September 11, 2001, and subsequent collapse of the World Trade Center Twin Towers and 7 World Trade Center, I know of no other high-rise building in the United States that has collapsed due to the effects of fire. The key to structural stability with any steel building is the proper fireproofing of the structural steel elements, and the subsequent proper maintenance of that fireproofing.

Ordinary construction. Buildings whose exterior load bearing walls are constructed of brick or concrete masonry units (CMU) are considered "ordinary construction" buildings. In general, brick or CMU can resist the effect of fires. The problem with ordinary constructed buildings is the floor and roof assemblies. They are often made of combustible or fire-susceptible materials. Ordinary constructed buildings with wood floors are prone to collapse if the floor assembly is weakened by the fire. Ordinary constructed buildings with steel floor or roof assemblies are susceptible to collapse if the roof or floor assembly is subjected to high heat conditions. It must be remembered in this case, high heat conditions are less than 1,000°F.

Another problem with ordinary construction is void spaces. These void spaces can be totally combustible, such as ceiling and joist (floor) spaces as well as interior wall assemblies.

Wood construction. Most of the fires we respond to are in wood structures. The exterior load bearing walls are made of wood assemblies. In most residential occupancies, these assemblies are wood studs (columns) of either 2×4-in. or 2×6-in. dimension lumber. The key in understanding the structural stability of a wood frame building is to know the approximate age of the building and how the exterior load bearing walls are constructed.

There are six types of wood frame construction. Three of the six are relatively common: balloon frame, platform, and truss frame construction. Log cabin, post and frame, and plank and beam are not as common except in a few areas of the country.

Balloon frame construction started in the 1830s. The studs run two or more stories high from the foundation to the eave lines. Balloon frame floor assemblies are tied to the studs by resting on a board laid horizontally called a ribbon. Buildings constructed in this manner are very susceptible to fire spread if fire enters the wall or floor assemblies.

Platform frame construction is built on a platform of 2×12-in. dimensional framing lumber that contains the floor joists of the first floor. Sub-flooring is then attached to the joists creating a platform on which the wall assemblies, both interior and exterior can be placed. The wall assemblies for the first floor are then built on that platform. Each wall assembly only goes as high as the ceiling of that particular floor, generally 8–10 ft. Another platform containing either floor joists for the attic or second floor of the building are built on top of the lower wall assembly. Once in place, wall assemblies are built upon that platform, and so on. This type of construction creates a natural fire stop with the platform for each floor of the assembly that is built. Platform construction began in the 1930s and 1940s and continues on to this day.

Truss frame construction is engineered construction in which the roof and floor assemblies and studs are tied into a unitized frame. The studs are an integral part of both the roof and floor assemblies. Truss frame is also called "lightweight wood construction" because the individual pieces of wood required are smaller than the typical 2×4-in. dimension framing lumber used in platform or balloon frame construction. Because of the design of the truss, these lighter, smaller pieces of wood can be used in its construction. With mass comes fire resistance. A 4×4-in. member is much more resistive to fire them 1×1-in. stud piece. That is why lightweight wood assemblies are extremely vulnerable to collapse. The connection that holds the engineered trusses together can be the weakest part of assembly itself. Connection failure is frequently a contributing factor in the failure of lightweight trusses exposed to heat and fire. The weakest part of any truss is the material that connects the web and the chords. Additionally, many "Glulam" arches, beams, girders, and columns are found in newer buildings. These construction members are an integral part of the system that holds up the building, floor, and wall assemblies, and are simply glued together under pressure.

There are also buildings that are made utilizing a combination of the above four methods. The bottom line is to know the buildings in your community and how they are constructed. Have a good understanding of the construction principles utilized. It is unfathomable to me that many firefighters today do not spend a significant amount of time learning about the construction methods of the buildings in their area.

Occupancy type

The occupancy type of a building has a lot to do with reading the building for search. Let me explain to you what I view as occupancy types. I define occupancy type as what the building is being used for. Buildings that were originally constructed as single family residential occupancies can have very different occupancy types. Of course they can be just a single-family residence or a house, as the name implies. However, houses have been converted into many different occupancy types or uses. Churches, group homes, restaurants, coffee shops, antique shops, and anything else you can think of are housed in what were originally single-family homes.

For the most part we can look at a building and from the outside get strong indications as to what the building's occupancy type may be. Churches look like churches. Grocery stores look like mercantile establishments. There are a few surprises out there when it comes to determining the occupancy type of a specific building. That is why it is important that if we see something that doesn't look right during our day-to-day activities, we stop and see what is going on.

Once you have a clear understanding of the occupancy type of a building, estimates can be made as to the number of possible victims and the priority that should be given to search. For example, vacant buildings are less likely to require a search. A single family home with cars in the driveway and bicycles on the front yard at 1:00 a.m. paints a completely different picture than the vacant building. I would expect to find victims in a house with bikes in the front yard. I would be very surprised to find victims in the vacant building.

The way in

The next factor to be considered in reading the building for search is the path into the building. There are several factors that need to be taken into consideration when the officer reads the building for a path of entry.

- **The placement of the initial hoseline.** It is safest and easiest if the search crew follows the initial hoseline into the building. First, it lessens the chances of the search crew being hit by the hose stream. Second, it gets a hose stream between the fire and the search crew. Lastly, the hoseline should be positioned so the nozzle is between the fire and savable victims. This should get the search crew to a point close to the fire where search efforts should begin.

- **The location of the fire as it relates to possible victims.** The point of entry should be considered in relation to where the fire and the victims could be. The intent is to keep the fire from getting between searchers and potential victims. A point of entry that keeps the victims between the fire and the searchers is generally most advantageous. In the absence of specific knowledge concerning the location of victims, it is best for the search crew to follow the attack line into the building. This provides the safest and most rapid path to the fire.

Locating stairs

One of the key factors in reading the building for search is locating the stairs in a multi-story building. Time is of the essence when conducting a search for victims in an occupied building. When you consider the time it takes for a person calling a fire in to realize there is a fire, locate it, and notify the fire department, and for the fire department to dispatch the apparatus, it leaves few precious minutes to find viable victims. If additional time has to be taken in zero visibility to locate the stairway to the second-floor once the search crew enters the building, then the margin for success is further reduced. Many traditional two-story homes provide an indication as to the location of the interior stairs. The location and size of some exterior windows, doorways, and the general style of house can provide clues as to the location of the stairway (fig. 12–1).

Fig. 12–1. Can you determine the location of the stairway from the outside on this house?

In one-story buildings, locating the probable area of victims from the outside can be increased by looking at the building. In a ranch house, sleeping areas are easily identifiable in relation to the location of the garage (fig. 12–2). In some commercial occupancies, the business or patron areas can generally be distinguished from the "workers only" areas (fig. 12–3).

The likelihood of victims

The likelihood of victims is generally a combination of several factors:

- **Occupancy types.** I would expect more potential victims in a restaurant on a Friday evening than I would at an auto repair garage at 3:00 a.m. I would also expect more victims to be endangered in a house fire at 3:00 a.m. then I would at 1:00 p.m.

Fig. 12–2. In most ranch houses, the sleeping areas are on the end of the house away from the garage.

Fig. 12–3. Picture of a one-story commercial store. Generally in these occupancies, "workers only" areas are in the rear of the building.

- **The intensity the fire.** There are several factors to consider, including the location of heat and smoke in relationship to the location of the victims. The human body can only tolerate 160°F dry heat and 130°F moist heat for a short amount of time.

- **The location of the fire as it relates to potential victims.** The lower the fire, the higher the problem. Conversely, the higher the fire is in the building, the lower the problem. Attic fires present much less of the threat to potential civilians inside a burning building then do basement fires. Hot gases, smoke, and heat all rise.

Where the victims might be

There is a bit of guesswork involved, but some generalities do apply. It is logical to believe that victims would be on the second floor of a two-story building during normal sleeping hours. It is also logical to assume that the same victims could be found in the first-floor living area during daylight hours. It also makes sense that no one would be in a dentist's office at 2:00 a.m. I always expect to have a severe life hazard in an occupied nursing home, no matter what time of day it is.

Where to Start and Where to Stop a Search

13

I don't remember anyone in the fire service ever telling me where to start and where to stop conducting a primary search. I went through some of my old notes that I had from drill school but I found nothing! I skimmed through the chapters of several of my old promotional exam textbooks and again found nothing.

Figuring out where to begin and end a search all goes back to common sense. There must be a logical place to start a search and as well a logical place to stop searching. It is the role of the incident commander to determine *when* a search will be assigned. I have provided specific cues to assist in making this decision. These are based on common sense and experience.

For the officer who has just been assigned to search, the first task in conducting a successful search and ensuring continuity of search is determining where to start your search. Almost everything else is subsequent to that first important decision. Even the decision on where to enter the building should be based on where you expect to start your search. Whether you enter in the front or the back by using doors, windows, or ground ladders should be determined in part by where you expect to start your search.

It is vital to pick the most appropriate location to start a search because of the ever-present time factor. Time is your enemy as it relates to search. Anything we can do to reduce the amount of time it takes to find viable victims increases our chances of success.

Recalling the partially involved house fire from chapter 10, I am positive that there are crews who would search the first floor in that house simply because they were there (fig. 13–1). I would assume that there is only a light haze of smoke on that first floor at this time. If a civilian were on that first floor, they would be in little immediate danger. That isn't to say we would not move the victim to the exterior of the house if I located one on the way to division two. Common sense should tell you that. But, considering only fire physics and human physiology, any civilian on that first floor is in little danger.

Fig. 13–1. Partially involved house fire. (Photo by Lt. Bob Pressler, FDNY [ret.])

My Rule

Normally search should start as close to the fire as possible where savable victims could be, then work out and up from that point.

Like much of the information in this book, this rule is based on common sense. If you look again at figure 13–1, you'll recognize that there could be no savable victims in the fire room in the B/C corner of division two. That room is in a state of flashover. Flashover occurs when the temperatures in the environment exceeds 700–1,400°F. There are only a few cold, hard facts concerning our job. One of those facts is that sometimes people die in fires.

Another hard fact is there still could be viable victims in portions of that second floor. If that is a typical, three-bedroom Victorian house with a bath on the second-floor, then we could have viable victims in the other two bedrooms and the bath area.

> **Scenario.** *The house in figure 13–1 is occupied by a mother, father, and two children (the floor plan is shown in figure 13–2). The parents' bedroom is located in the AB corner of division two. One child's bedroom is in the BC corner which is involved in flashover. The other child's bedroom is in the CD corner with the bathroom located in the AD portion of division two.*
>
> *The fire occurs early on a Sunday morning and the parents are both awake and downstairs having a cup of coffee. The children are both still asleep in their bedrooms.*
>
> *The child in the fire room in the BC corner of division two is dead. The child in the bedroom and CD corner of division two may still be alive but time is of the essence.*

I would be very upset if I lost both children in the above scenario. We're drawn like a moth to flame to the flashover room, but the 30 or 45 seconds we spent checking the dead child's room could have been used removing the living child. That 30 to 45 seconds may have made the difference in the life of that second child.

One exception to the rule

I believe in never saying never. I also believe in *never saying always*. Rules of thumb are good tools, but there are exceptions to almost every rule. There is one exception to my rule of where to start search.

Search should start as close as possible to the fire where savable victims could be and then work out and up from that point. The exception to this rule is when victims are believed to be located in an area remote from the fire. This exception has two common applications in our bread-and-butter, single family residential fires. The first exception is when the fire is all on division one and possible victims could be in sleeping areas located on division two. This scenario is most likely during a nighttime or early morning fire.

The second exception would be in a ranch style home where the fire is located in the living or cooking area and the victims are in sleeping areas. In these instances, search should commence in an area where the victims are believed to be. You should still start close to the fire and then proceed out and away; just do so from the most logical point you'd find savable victims (fig. 13–2).

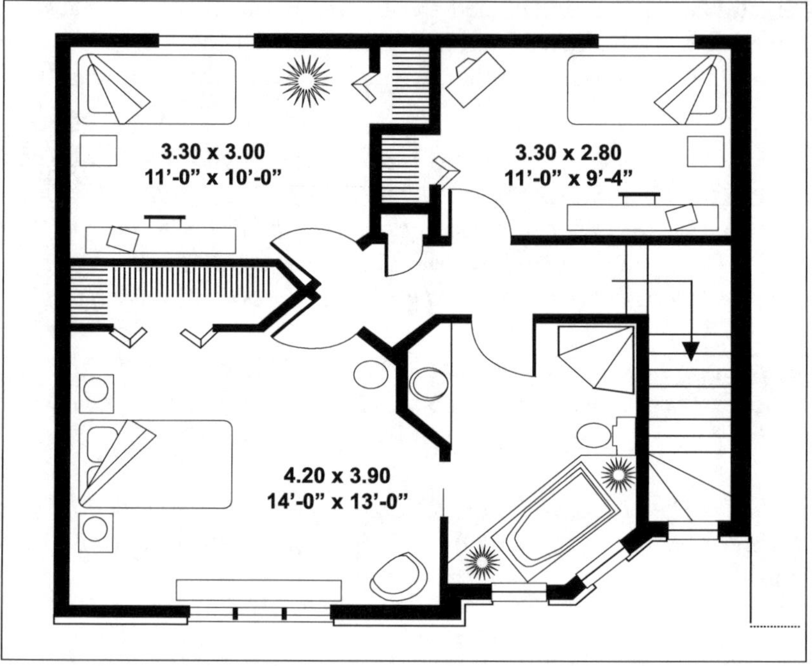

Fig. 13–2. When fire at 3:00 am originates in the living area of this house near the dining room, search would begin in the sleeping areas. This is an example of the exception to the rule. (Copyright Jim Barna Log Homes, courtesy of The HousePlanShop, LLC)

Where to Start Search In Detail

Search should start as close to the fire is possible where savable victims could be found. In a routine fire, attack is assigned first. With the exception of "seen" or "heard" victims or severe exposure problems, the generally accepted practice is to put the fire out first. Absent of the above two reasons, I can think of only a few fires in my 32-year career where this practice wasn't followed.

Following this practice does several things. First, putting out the fire eliminates most of our other problems. Smoke will still be present in the fire area and the fire will need to be overhauled, but the quicker we put out the fire, the less pressing our other problems become. Second, this gets a hoseline positioned in the most appropriate location. Hoselines should be initially placed with the intent of cutting off the spread of fire. Get a hoseline between fire and savable victims or savable property. Positioning a hoseline first also gets a hoseline between the fire and the search crew.

The hardest part of performing the rule on where to search first is in determining where savable victims could be. There are several indicators that will assist you in determining an appropriate place to start your search.

- **Heat and fire conditions.** I advocate that the search officer rapidly and briefly expose a small area of skin to determine the amount of heat in the area. Short of some form of thermal indicator, this gives you the best indication of the amount of heat in the environment. If you are experiencing a rapid buildup of heat coming from your bunker gear, conditions maybe too hot already. This exposure of skin can also indicate the survivability for unprotected civilians. My oven at home has a "warm and hold" setting that remains at a constant 170°F. I can stick my bare hand in the oven for short periods of time with little or no discomfort. At 170°F we are approaching the point where survivability of the unprotected civilian diminishes greatly.

- **The condition of the area where you begin search also gives an indication as to the survivability for unprotected civilians.** Heavy char in the area would indicate rapid, intense burning. If the char is low on the walls then the likelihood of survivability is diminished due to high heat concentrations close to the floor. The search officer should evaluate heat and fire conditions as they approach the area to be searched. The officer should then make a judgment call based on experience and physical factors to determine areas of likely survivability for civilians.

- **The amount of fire damage.** Another indicator used in determining a search starting point is the amount of fire damage in the area. Furniture and other household items that are heavily damaged by fire indicate high heat conditions and diminish the survivability of unprotected civilians. This is especially true of rooms involved in flashover that have been knocked down quickly, prohibiting the charring of wood products but facilitating heat destruction to other materials in the room.

- **Smoke conditions.** Physiology plays an important factor here. Every firefighter should be aware of the toxic effects of smoke. They should also know about the ability of the by-products of combustion to displace oxygen with other deadly gases, including carbon monoxide. Carbon monoxide in high enough quantities can cause death within minutes.

Covering the Initial Area of Search

Once the search officer has determined the appropriate area in which to begin the search, then the search should progress out and away from the fire. I say this because of the fact that the people in the most danger are closest to the fire. Cover those areas first, then work out and away from the fire until you either cover the remainder of the area on that floor or until the conditions are no longer IDLH.

There are no hard-and-fast rules in determining when an area is no longer IDLH. If you believe that there are still savable victims further in the building, then you can move to areas that are more highly IDLH. If you are experiencing 10–15 ft. of visibility and little heat, then you should probably move to a more IDLH area and cover it first, returning later to the less dangerous rooms.

If there are multiple floors or levels in the structure being searched, then once the initial level of the building has been covered, the search officer should attempt to search above the fire for victims. If the building is a two-story home with an attic, then *the attic should be searched any time there is a permanently affixed stairway.* This should include a pull down attic stairway *only* if it was actually pulled down and ready for climbing.

Once a searcher gets to the top of the attic stairway, a quick sweep should give an indication as to whether the space is used or not. If your first sweep determines it's an un-floored attic, then I would not search any further. If during your initial sweep the searcher finds Christmas trees, boxes, and little room to move about, then I would not expect civilians to be in that space. If the sweep locates a bed, chair, or dresser then I would expect that area to be searched.

If you're searching an occupancy other than a single-family house and there are additional levels above the initial search area, you should cover the area above the initial search area first. Continue until you cover the entire floor level or the area is no longer IDLH. If you cover the whole floor, you should move up to the next level. Continue this process until you reach the top floor or living space in the building.

At that point you should go back to levels previously covered. If there were areas you did not search because they were no longer IDLH, take a quick look (because of improved non-IDLH visibility) through those areas for victims. After those areas are covered, you can start your search in reverse order below the fire until you run out of area or IDLH.

At that point in time the officer should report to command that the officer has an "all clear." Then they should inform the incident commander that the crew is either low on air and coming out or has sufficient air to start another task. That decision should be left to command.

How Long Should a Search Last?

14

There is a time frame beyond which search should not be conducted any longer. There are several common-sense reasons for this. However, this is one thing I was never taught, nor have I read about it in any fire department texts.

There are some human physical factors along with common sense that apply here.

Time Frame

Under IDLH conditions, a primary search should be conducted for no more than approximately 15 minutes. There are some human physical factors and common sense that apply here.

The duration of an SCBA bottle is an easy way to track the amount of time being spent on search. It is the officer's responsibility and is part of the practice of air management. Generally, fire departments use 30-minute SCBA bottles. As we all know, the 30-minute designation is a misnomer. Very few firefighters can make the air supply contained in a 30-minute SCBA bottle last for 30 minutes. In tests conducted in Toledo in 1984, the average air consumption of a full 30-minute SCBA bottle was closer to 16 minutes. Those 16 minutes are the time from the first breath on a full SCBA bottle to the last inhalation. Using proper air management techniques, we should not be in a building utilizing SCBA to the point that our low-air alarm begins to sound.

Proper air management techniques should be utilized at any working fire. This is especially true of commercial occupancies. If used, then most crews will have approximately 10 to 15 minutes in IDLH conditions prior to the sounding of the low-air alarm and the crew exiting the building. A judgment call must be made following individual department guidelines in the event of a life-and-death search. Your department should have rules in place that define when crews are required to finally abandon the search due to low air.

Humans can only survive with 15% O_2 or less for four to six minutes at temperatures of less than 180ºF. We should give all victims the benefit of the doubt and continue our search for that 10-to 15-minute duration of a full SCBA bottle.

This is where common sense must be applied. These two facts—the standard SCBA bottle has a 15 minute average supply and the human body can only survive for four to six minutes in less than 15% oxygen—are givens.

We could require search teams to utilize larger capacity one-hour SCBA bottles. Even if one-hour bottles became the industry standard, it still wouldn't negate the human survivability factor. Let's the look at the following two scenarios:

> **Scenario 14–1.** *We have a working fire in a three bedroom, one bath, two-story home. The fire occurs at 3:00 a.m. in the morning with heavy IDLH conditions throughout the first and second floor. A crew of three firefighters and an officer are assigned to search. From the time the crew goes "on air" at the front porch, it takes them 7½ minutes to search the entire first and second floor. In my opinion, this is an acceptable standard practice and certainly could be defended in any court of law.*

> **Scenario 14–2.** *We have a working fire in the lowest level of a typical, three story, garden-type apartment building containing 12 separate units. Heavy IDLH conditions are present on all three floors and in all 12 apartments. A crew of three firefighters and an officer are assigned to search. From the time the crew goes "on air" at the front door of the lowest level, it will take them approximately one hour and eight minutes to conduct the search. This extensive search will require three SCBA bottle changes, as well. The crew is trained to conduct a bottle change in two minutes or less. Altogether more than 6 minutes will be used simply to replenish their air supply, not including travel time from the structure to the engine and back. More than 50 minutes were spent in actual searching. This is not an acceptable standard practice. If this fire produced civilian fatalities, I believe the incident commander and department could be exposed to lawsuits for negligence.*

Search officers gauge search with maximum time frame in mind

It is the responsibility of the incident commander to assign the appropriate number of crews to successfully complete a task. However, the individual search officer also has a responsibility to keep the incident commander apprised of the progress of the search. There are certain situations where an incident commander can believe the appropriate number of firefighters are assigned to conduct a search. But if the search officer believes that the assigned crew cannot physically cover the area assigned in the allotted 10 to 15 minute time frame, then it is the search officer's responsibility to make the incident commander aware of this fact.

I've seen this issue arise before in a single family residential occupancy that appears to be typical from the exterior. Upon entry, the search officer realizes the occupant is a pack rat and the house is very tightly cluttered with every imaginable type of object. The obstacles created by the resident's hoarding dramatically slow the search effort. Under these conditions, the search officer should notify the incident commander that search will not be able to cover the entire house in one bottle change. It is then up to the incident commander to take whatever steps deemed appropriate.

Know the Ability of Your Crew

The above scenarios discussed single family residential occupancies of relatively standard proportions. Many of us have single family homes larger than 6,000 sq. ft. in our response districts (figs. 14–1 and 14–2). How many square feet can your four-member crew cover? Let's do some math! If you expect to completely search a 6,000-sq.-ft. home in 15 minutes, your crew will have to cover approximately 400 sq. ft. per minute. Assuming that the officer is not actually participating in the search, that means each searcher must cover approximately 150 square feet every minute. That is the size of a 10×15-ft. bedroom. Do you believe your crew could accomplish this?

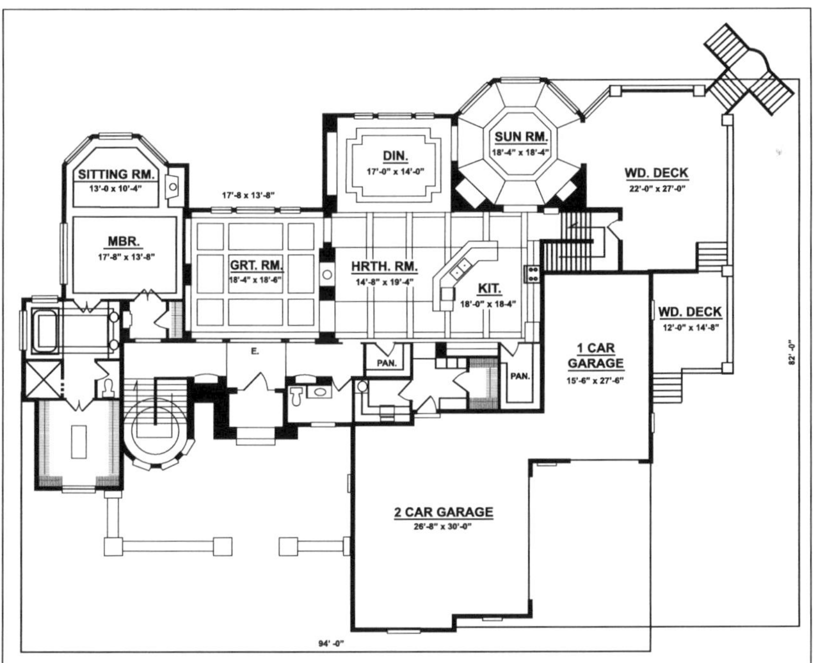

Fig. 14–1. This is a floor plan of the first floor plan of a very large home. (Copyright Ahmann Design, Inc., courtesy of The HousePlanShop, LLC)

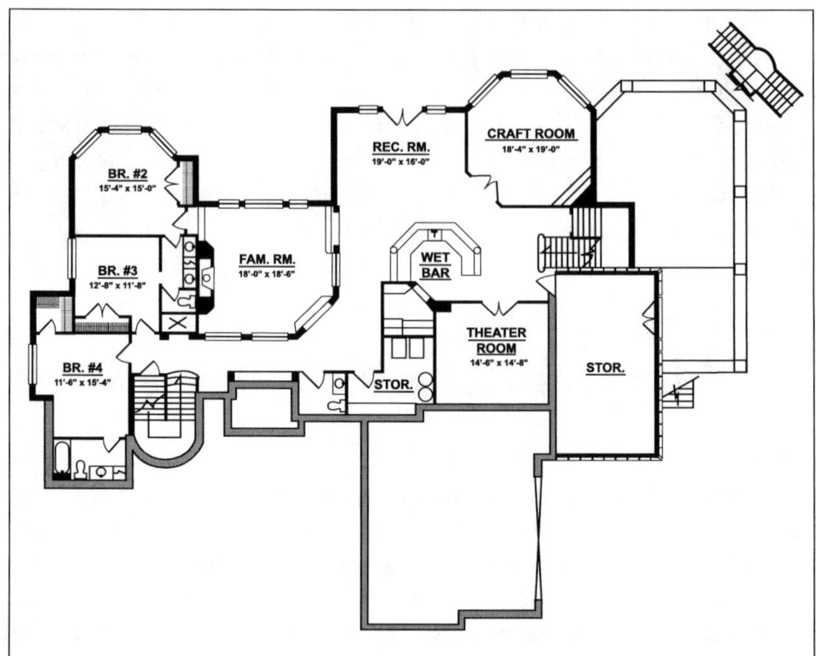

Fig. 14–2. This is the second floor of the same house. (Copyright Ahmann Design, Inc., courtesy of The HousePlanShop, LLC)

Now let's look at the garden apartment scenario. Let's assume one officer and three firefighters make up the search crew and they have the standard 15-minute SCBA time constraints. The crew would have to complete each apartment in 1 minute and 15 seconds. If each individual searcher covered a specific apartment by himself, they would be required to complete each individual apartment in 3 minutes and 45 seconds in order to stay under the maximum allowable time frame. The average two-bedroom, garden-type apartment in my community is approximately 900 sq. ft. In order to accomplish this search, each individual firefighter would have to cover approximately 240 square feet per minute. Do you believe your crews can accomplish this?

Up until this point we have not discussed commercial occupancies. A single crew consisting of three firefighters and an officer could search an average auto repair garage in about 15 minutes. There are many commercial occupancies that can be adequately searched by a single crew in an acceptable amount of time.

However, there is also an extremely large number of commercial buildings in any response district that cannot be covered in less than 15 minutes. Grocery stores, drugstores, other mercantile outlets, larger family chain restaurants, and large manufacturing plants are among these massive structures.

It is incumbent on every company officer and incident commander to fully understand the capabilities of the crews under their command. This not only applies to search capabilities, but also to the amount of water an engine crew can safely flow from hand lines, as well as the time required to establish that flow. They must also know the amount of time required for a truck crew to vertically vent the roof of a commercial, metal deck build-up rough assembly. Not to do so borders on negligence. In today's litigious society, if you don't know the abilities of your crews then you're just begging for long, complicated court battles.

Although this may not be the time or venue to discuss this, I believe I would be negligent if I didn't take a little time to brag about my former department. Toledo is one of the few accredited metropolitan fire departments in the United States. Being one of the members of the accreditation team and, at one time, the accreditation manager, I am extremely familiar with the process of being accredited.

In Toledo we required officers to provide benchmarks on every assignment they were given. These were standard across the department and, when given, were time stamped in the computer aided dispatch (CAD) system. As operations deputy, I would review the data gathered from every working fire. We defined a working fire as any fire which utilized three or more apparatus for longer than 30 minutes. A typical garage fire that required two engines but took an hour to bring under control and overhaul was not considered a working fire.

We learned many things from looking at this data. We could compare the benchmarks between different occupancy types that were listed on the National Fire Incident Report (NFIR). For example, we looked the benchmark for search in apartment buildings, which was "all clear." In the case of apartment buildings similar to garden type apartments, we learned that a single crew could search about six apartments prior to running out of air using a 10- to 15-minute time frame. This told our incident commanders that it would take two crews to cover 12 apartments in a single section of garden apartment building while still remaining under the 10- to 15-minute time frame. Because of this reason, we sent an extra apparatus to any fire in apartment building of that type.

We looked at time frames concerning fire attacks and ventilation as well, which gave our battalion chiefs a very educated understanding of what crews in Toledo could accomplish at a typical fire. We used this time reference as a training tool in the event that a particular crew was not meeting the departments average evolution time. That information proved to be very helpful to us on several different levels, as it could for your department.

Searching with a Hoseline 15

So far I may have said a few things that piqued your interest and some other things that you flat out disagree with. That's okay. The topics I discuss in these next two chapters are among the most controversial in this book. Even so, these are two of the areas about which I am most adamant. Remember the question I posed in chapter 1? Are you searching the way you would want someone to look for you or one of your family members?

I believe the officer should do everything possible to simplify the search. The same holds true for most of the evolutions we conduct in the fire service. Evolutions used to mitigate a hazmat incident are going to be relatively complicated, to be sure. That's all the more reason to keep them as simple and straightforward as possible. The same issues are relevant at trench rescue incidents. These are technical rescues by their very definition. We call them that for a very good reason. I still want the pulling and stretching of a hoseline, raising of ground ladders, and conducting searches to be as uncomplicated as possible. By doing so we save time and effort, both of which are priceless commodities when you're talking about fire suppression and search.

Searching With a Hoseline

When I teach courses and give lectures, the subject of search always comes up. When I discuss my thoughts on searching with a hoseline, I get a lot of raised eyebrows. After a thorough explanation, I usually don't get much resistance to the concepts. I remember teaching these concepts to my own department's company officers several years ago. When I presented the slide concerning searching with a hoseline and gave my explanation, I got some very heated discussion from a few of our company officers. They were extremely concerned, especially when we discussed going above the fire. For example, imagine a nighttime fire involving a two-story single-family home. The bulk of the fire is located on the first floor, but the search team isn't taking a hoseline with them.

I simply asked them to cite for me the last time they pulled a hoseline with them when they were assigned search. I asked them when was the last time they remember seeing anyone in Toledo take a hoseline into a residential occupancy with them while they conducted a search. The room got quiet. A few smiles started to spring up, then I moved on with our discussion. It sounds like it goes against basic safety principles, but upon reflection, we simply did not do it!

Entering a burning building without taking a hoseline for protection goes against logic, common sense, and many safety principles. Taking a line with you is what you've been taught. It's what we've come to expect as the norm, but when was the last time you took a hoseline with you to conduct a search?

I'll come right out and say it. I would *almost* never expect a hoseline to be pulled when conducting a search in a single family residential occupancy. I said *almost* never. I suppose if I sat and thought about it for a very long time, I might come up with a reason to pull hoseline for search. Some of the 6,000-sq.-ft. and larger homes that are being built today may require the use of a hoseline. They don't need it for protection from fire, but as a method to remain oriented with your means of egress.

In the next few paragraphs, I deal predominately with bread-and-butter fires in single family residential occupancies. I will go into detail about less common occupancy types later in the book.

Two reasons to leave the hoseline outside

Many officers are adamant that a search crew must have a hoseline with them at all times. I would imagine the most common reason they cite is protection of the crew. I have two responses:

- If you don't feel comfortable searching an area due to heat and fire conditions, are you searching where savable victims could be found? Our bunker gear and other personal protective equipment allows us to enter environments that the unprotected civilian cannot survive. Even our PPE has its limits, approximately 500°F for up to 2 minutes, depending upon how clean the bunker gear is. 500°F is not yet hot enough to support flashover. If you are becoming uncomfortable due to heat, you're probably in too hot an environment already. If you don't feel comfortable, can there be any unprotected civilians still alive in that area? The answer is, "No. You're looking in the wrong place! You're looking where only dead people could be. You're wasting precious time that should be spent on victims who may still be in the building, but in environments more conducive to supporting life.

- Staffing is an issue for today's fire service. Even the larger departments like the FDNY are reducing staffing. The average number of firefighters on a company today is less than four. Even assuming that four is the number of members of every company in your department, let me pose this question to you. How many firefighters does it take to pull, stretch, and *maintain* (keep it with you) a 1¾-in. handline? Two? One? To pull and stretch is one thing. Climbing stairs, pulling the line around the stairway landing, up the next section of stairs, and then around a hallway requires at least two firefighters. There's no sense pulling and stretching a line while conducting the search if you don't keep it with you. That means you have two firefighters whose focus is directed on the hoseline and not the search. You've effectively cut your search crew in half. If you have a four-person crew, that only leaves two to search (and one of them is the officer—who should be directing, not searching). Is that the way you want a search crew to look for you or your family members?

Some Givens

In order for any search to be conducted as safely as possible in the absence of a hoseline, there are a few givens that must be assumed:

- Prior to commencing search, an attack crew has been assigned and is moving toward the fire. The fire may not be darkened down yet, but a hoseline is being stretched into the building. That will put water between the fire and savable victims, savable property, and the search crew.

- The search crew took the proper way into the building by following the attack line. This route most often goes through the front door of a single-family home. That line is cutting off the spread of fire from savable victims, property, and the search crew. Following the hoseline lessens the likelihood of the search team being hit by a hose stream. If search follows the line, they should be taking a path that leads them to the fire. Search should start as close to the fire as possible where savable victims could be. If the search crew follows the hoseline in through the door and moves up to division two, attack should have a line already positioned to protect them from the spread of fire. In the event there is fire on divisions one and two with savable victims believed to be on the second floor, the search crew should delay moving up to division two until the fire on division one has been darkened. Then search should follow attack up to division two and begin search efforts. Without an adequate number of crews to pull and stretch simultaneous hoselines, I can think of no other way to safely darken a fire and commence a search in those conditions. Oriented VES could be used from the outside, but only in extreme rare conditions where viable victims are almost assuredly present.

- The search officer is doing his or her job. The first priority of a search officer is the safety of the crew. Search is a tough, dangerous assignment. Searches where there is no IDLH or likely victims can be delayed for other priorities. Searches with life hazards and victims require more expediency. Drastic times call for drastic measures. It is in these drastic times when it is imperative that the search officer's focus stay on crew safety and fire conditions in the immediate area. Sampling the environment for heat conditions and communicating with the attack officer are two requirements for the search officer, especially when the officer and crew are above the fire without a protective hoseline.

Searching with a Hose in Commercial Occupancies

I do not advocate for the use of a hoseline when a crew searches a residential occupancy because it slows down the search. However, I do recommend the use of a hoseline when searching most commercial occupancies. The hoseline is pulled and stretched, but not used as a tool to fight fire or defend the crew. Instead, it is used as a 2½-in. wall with which the oriented member maintains an awareness of the egress. It is much easier to find a hoseline in zero visibility than it is to find a piece of ⅜-in. rope.

Searching with a Tool

16

There are just as many instances of "always" in the fire service as there are "nevers," which is to say, not many. I may go out on a limb in what I advocate in this chapter. I have discussed these methods with my fellow firefighters over many a glass of Irish whiskey. Some of those discussions got pretty heated—hopefully not the fault of the whiskey!

Repetition is one of the most effective ways to learn, so please forgive me if I sound like a broken record. I believe a major flaw in the fire service is that firefighters today are not reading the books that older firefighters trained on, and vice versa. I was shocked when I read excerpts from one of the premier fire recruit guides concerning many aspects of our profession. They said things contrary to what I was taught and had used successfully throughout my entire career.

One such instruction was, "Never vent a roof without taking a hoseline with you." First off, I never say never. Second, having been assigned to a truck company for several years, I went up many roofs with an ax in hand, intent on venting the roof. This was in the days before carbide-tipped chain saws and positive pressure fans.

I do not remember one instance where a chief, an officer, or I have ordered that a hoseline be taken up in order to successfully vent. For one thing, it would only slow us down. For another, a hoseline on a gabled roof is a very dangerous tool to operate and maintain. Also, the best thing a truck can see when venting a roof is flame coming from the vent hole. A truckie with a hoseline in hand and flame pouring from a vent hole is going to shove that stream right down the vent hole, which is the exact wrong thing to do.

I simply do not understand the thought process. If your intention in taking a line to the roof is to provide attack capabilities, you do not know the first thing about physics, ventilation processes, or the moth-to-candle syndrome. If your intention is to keep your crew from being cut off from their means of egress, then your officer is positioned on the wrong side of the vent hole. If you intend to use the hoseline as a secondary means of egress should you be cut off by fire then a) the officer is not doing his or her job and b) you may be an idiot.

But let's get back to search.

Another statement in this fire recruit guide was that firefighters should never enter a building without a tool in their hand. Again, there's the word *never*! Under *most* circumstances, I believe a firefighter should not enter a building without a tool of some sort in hand, be it an axe, a Halligan or the nozzle of a hoseline. Truckies should not go onto a roof intending to vent without every member having a tool, be it an axe, a pry bar or pike pole, or a carbide-tipped chain saw.

Should a Tool Be Used in Search?

The one exception to a firefighter entering the building with a tool to conduct an evolution is when the evolution is search. I do not believe it is in the searcher's or victim's best interest for the searcher to carry a hand tool, as is standard practice in many departments across United States today. *It dramatically slows down the search!*

I will illustrate my point with an example—this will require you to visualize what I'm saying. Please do so with an open mind. This is how we are taught to conduct a one-person search or use the standard method of search.

Our firefighter enters the room. When searching with a tool, one of the first preparations is to put the tool in the proper hand and position. The proper tool position for search generally depends on the tool. For example, when using an axe, the searcher should carry the axe blade in hand and use the handle to sweep with.

After he positions the tool, our firefighter commences with the search pattern. He crawls one or two crawls forward along his wall, then side-craws laterally toward the center of the room and sweeps with the tool. If nothing is found, he will move back to the wall, then again crawl two or three crawls forward along his wall and repeat the process.

This time, when moving laterally toward the center of the room and sweeping, he contacts something with the axe handle. Because there is no feeling in the end of the axe handle, he moves laterally toward the object the axe handle hit, then drops the axe and uses his tool hand to feel the object. (Remember I said "never" turn your body so that your shoulder is perpendicular to "your wall" when doing a one-person search. Doing so can get you lost and un-oriented to your wall. So you must in this case drop the tool to feel the object.) In this instance, the searcher determines the axe hit the side of an overstuffed couch.

Now he must pick the tool back up and position it correctly again before continuing the search. He crawls laterally back to his wall, then crawls forward, side-crawls toward the center of the room and sweeps again. This time as he sweeps the axe toward the center of the room, he gets the axe handle tangled in the legs of a table (fig. 16–1a and b). This hampers his ability to cover the entire area on either side of the table and he can't tell if he has covered the far side of the table. After dropping the axe and feeling on all four sides of the coffee table, he attempts to find his axe, which he laid on the side of the table closest to the wall. After several seconds, he finally locates it, repositions it again, and moves back laterally to his wall. A third time he hits something with the axe handle. He crawls laterally to find the object, drops the tool, and works to identify the object by feel. It's a pillow lying on the floor.

Fig. 16–1a & b. How long would it take you and how hard would it be to sweep under either of these tables with an axe or Halligan?

Come on! Is that the way you would want me to look for your child or another loved one? I don't believe so. Using a tool to assist in searching only slows down the search. In my opinion, it does nothing to improve the speed, proficiency, or any other quality of the search.

Why Carry a Tool?

I can only think of two reasons a searcher would even contemplate carrying a tool while conducting a search:

- *To extend the reach when sweeping toward the center of the room.* To this, my answer is to take one more crawl laterally toward the center of the room, then sweep with your hand. In conducting this simple act, you compensate for the length of that tool. Also, if you use your gloved hand rather than a tool, you don't have to drop anything to feel the object you hit, or reposition anything before continuing with your search. You simply sweep, hit something, feel and identify it, and if it is not a victim, move on—saving an unbelievable amount of time.

- *To assist in the event you get in trouble during search and need to breach a wall or otherwise break something to save yourself.* Whenever I teach the subject of search, this almost always (noticed I didn't say *always*) comes up. When I discuss the two previous points I stop and ask everyone in the room to raise their hands if they ever had to use a tool to save themselves while conducting a search. I have asked this question to tens of thousands of firefighters. Only once did a person raise his hand indicating that he had indeed had to use a tool in order to save his life while conducting a search. I asked him to provide us with the circumstances of the incident and after listening, I believe he may have confused some of his facts. I'm not saying he was not telling the truth, but the way he conveyed the story to me and others in the room created some doubt in my mind.

Who Should Carry a Tool During Search?

Every time a fire crew conducts a *bona fide* search in IDLH conditions, they put themselves in harm's way. In previous chapters I have stated that I do not believe the search crew should take a hoseline with them in residential occupancies. Having said that, it is imperative that the search crew take a tool. I believe there are two options available—one is a contingency and the other is somewhat a given.

The contingency is that each searcher can take a tool into the building up to the point of starting the individual primary search. For example, a searcher brings in a tool and takes it to the first room that the oriented officer has directed to be searched. At that point the searcher leaves the tool and enters the room to commence with the search. My recommendation is that if you conduct a left-hand search, place a tool on the right side of the door that you entered. This way, when you reach your tool you will know you have covered the room and your door to exit is right there. Leaving it there also lessens the chance that you'll forget it as you exit the room.

Another version of contingency is to stick the tool in what in Toledo we called a *hose belt,* so it is with you in the event of an emergency, but you do not use it to sweep for victims.

A last version is for the searcher to keep the tool in the hand closest to the wall and not in the hand the searcher is sweeping with.

The given is that the search officer should always have a tool in the event the officer or a crew member gets in trouble. If conducting a standard search, all members are participating in the search together. In this case, the officer keeps the tool in the hose belt or in the hand closest to the wall to avoid the time-consuming act of using the tool to sweep. If using the oriented method, the officer should keep the tool while moving from area to area inside the building. Remember, the officer must never (that's a real *never*) be too far away to communicate effectively with every member of the search crew. This normally keeps the officer within 20 feet or so of every member of the crew. In doing so, no firefighter is ever any more than 20 feet or so from a tool.

If you doubt what I have discussed in this chapter, I ask you to do one thing prior to making your final decision. You should actually watch a firefighter conduct a search using a tool to sweep with. My suggestion is to get a recruit firefighter bunkered up in full turnout gear with a blacked-out facepiece. Give the firefighter a tool and the instruction to search the dormitory or kitchen in the fire station, or the training tower, or somewhere where there is some furniture to encounter. Don't mess with the searcher, but rather encourage him or her to do the best possible job—then just stand back and watch the search. Ask yourself if that is the way you would want someone to look for you or your family members? I believe you'll realize the answer is *no.*

Searching with a Thermal Imager

17

A few years back in the Round Table column in *Fire Engineering* magazine, the question was asked, "What do you think is the most significant advancement or invention in the fire service?" I got several good answers to that question. Pumps, hose, self contained breathing apparatus, radios, and thermal imagers were among some of the most frequent responses.

Thermal imaging devices are relatively new tools in the fire service arsenal. Most departments have a limited number of imagers available to them. Although the price of an individual imaging device has dropped over the last few years, it's still a very expensive tool.

As it relates to search, the thermal imager provides the user the ability to actually look for victims instead of blindly feeling for them in heavy smoke IDLH conditions. This is truly an advantage and, in some instances, speeds up the search considerably.

Unfortunately thermal imagers have a very serious limitation in conducting a search. Generally an entire crew has only one imager with which to work. Most departments require the officer or another firefighter to use the imager. They direct the remaining crew members to follow by physically contacting the member with the imager while moving throughout the structure being searched.

When a specific area is to be searched, generally the member with the imager scans the area rapidly and then paints a very brief picture of the layout of the room to the member who will conduct the actual search. The member is told to do a left- or right-handed search, usually to a specific area that is out of sight of the imager. The searching member covers that area and then moves back to the point of entry, thereby covering the entire area, including the areas that the imager could not see.

If scanning a bedroom, for instance, the officer will scan the room rapidly, noting the position of the bed and any other large obstructions to the view. The officer should also quickly scan the top of the bed to make sure no victim is there. The officer will then instruct the firefighter to do a left- or right-handed search, depending on which route provides the quickest access to the unseen area on the other side of the bed. The firefighter needs to move along the wall, skipping the normal side sweeps toward the center of the room because that area has already been "cleared" using the imager. Upon reaching the bed, the firefighter will quickly search the unseen area as instructed by the officer and sweep under the bed as rapidly as possible. If the officer or firefighter notices a closet door, the firefighter should rapidly cover that closet area as well. The firefighter can then move back as fast as possible to the door and the crew will continue on with the search.

Limitations of the imager

There is one major problem when using a thermal imager as it relates to search. That problem will be totally rectified when every member has a thermal imager that, ideally, will be built into the lens of the SCBA facepiece. Absent that, they are not the panacea to search that some would have you believe. The problem with using a single thermal imager while conducting search doesn't present itself while the search is being conducted, but instead develops upon the location of a victim. It is at this point the stumbling block arises. Let me present you with a scenario to illustrate my concern:

Scenario 17–1

A fire crew consisting of an officer and three firefighters is assigned search in a single family residential ranch-type house. The house is approximately 5,000 sq. ft. One imager is available to the crew to utilize during the search. As dictated by procedure, the officer is the member that uses the imager.

The crew dons full safety equipment and enters the house. Because of the time of the fire, the officer heads for the sleeping areas located on the B side of the house. The officer determines this by the location of the garage on the D side of the house and the arrangement and size of windows on the B side. As the officer enters the building, the crew members follow her in, each one maintaining physical contact with the preceding firefighter. The officer leads them to the area to be searched and finds a hallway leading to all four bedrooms. As the crew enters the house, they do not pay particular attention to how they entered and what they passed, information that could be used later to reference their means of egress from the building. This tends to be a typical process when using an imager.

The officer instructs a firefighter to search the first bedroom they come upon. The officer scans the room and gives the firefighter his direction of search and other information pertaining to the room. As the firefighter enters, the officer scans the hallway and sees another door in a very close proximity to the area where the first member is currently searching. The officer scans that second room and sends the other two firefighters in to begin a search. The firefighter in the first room informs the officer he found the child under the bed.

At this point, all firefighters must leave the building because they have no way of knowing where they are and where their path of egress is located. The officer can't give the imager to the firefighter who located the child so that he could take the child outside, because that would leave three firefighters inside a building and only one of them knows where the egress point is found. That egress point is locked in memory by visual cues, not "felt along" cues. This poses a very difficult perspective in locating your means of egress. The only safe thing to do is to have the entire crew stop the search and remove the located victim.

The fact that only one imager is available prohibits continuing on with the search if a victim is located. The above scenario is in a relatively simple single-family ranch house. Envision the difficulties that could present themselves in the same type of scenario, but located in an occupancy such as a restaurant, nightclub, or mercantile establishment.

Again, an imager is a wonderful tool. I believe they will save countless lives when more than one is available to a specific crew conducting a search. Two imagers per crew of four would allow two firefighters to remove a located victim while two other firefighters with another imager continue on with a search.

If you only have one per crew, I am of the opinion that the officer should be the member utilizing the single imager assigned to a crew.

Single–Family Occupancies 18

Aside from downtrodden people who find themselves homeless, most of us live in some type of structure. Many of us, especially those living in metropolitan areas, live in multi-family residential occupancies of some sort. The vast majority of families, however, live in single-family residential occupancies. Although statistically there are other occupancy types that kill firefighters with more frequency, single-family occupancies record the highest annual number of fires in the United States. They also lead in the number of yearly civilian fatalities in the United States. These are truly the "bread and butter" fires facing today's fire service.

Even though I don't discuss much building construction in this book, I find it unimaginable that a firefighter today would crawl inside a building without a solid understanding of how that building will react to the affects of fire. I would be negligent if I did not provide at least a few basic building construction fundamentals. If we are to be successful at locating and removing civilians endangered inside a burning building, we must have an understanding of the effects of the fire on the materials that are holding the structure up over our heads.

Problems Searching
Single–Family Occupancies

Single-family occupancies are truly the bread and butter fire for the majority of departments in the United States. More civilians are killed and injured in residential fires than any other occupancy type. As often as we see these fires, you would think the death and injury rates that result would drop over time. Instead, we're still losing approximately 4,000 civilians every year in single-family occupancy fires.

Some of the problems associated with single-family occupancies are:

- **An unknown number of potential victims.** There is no building code that I'm aware of that dictates or prohibits the number of people that can be allowed in a single-family occupancy.

- **Limited fire codes.** There are very few fire codes that restrict what can and cannot be done inside a home. As is often said, "A man's home is his castle." For example, barbecue grills are prohibited within 10 feet of multiple-family dwellings. This code does not affect single-family dwellings. Flammable liquids must be stored in approved lockers in most occupancy types other than single-family occupancies. You can store gasoline anywhere you like in your home.

- **Varied layouts.** Layout inside these occupancies is rarely restricted by any code. We can pre-plan commercial occupancies, but we cannot pre-plan each individual house. No department has the staffing or resources to pre-plan, document, and maintain a retrieval system for every single house in their area. We rarely enter a specific house prior to a fire. Some homes are visited by the fire department for emergency medical runs, but focus on these instances is patient care and not any form of pre-planning. We don't roam throughout the entire house on an EMS call, as is required during a formal pre-plan.

- **Vacancy.** It is often difficult to determine if victims are actually inside a single-family dwelling. You can generally be assured that there is no human being inside a barbershop at 3:00 a.m. The same can't be said for a house. At times it can be difficult to determine if a house is even vacant or occupied.

- **Sleeping victims.** I expect that victims could be asleep and in a lowered state of alertness at any time of day in a home. Although many jurisdictions require smoke detectors and other early warning devices in homes, enforcement of those codes is very difficult. Early notification systems may not be available to occupants in a lowered state of alertness.

Building Construction 101

Most single-family residential occupancies in the United States today are some form of wood construction. The exterior load-bearing walls that are holding the building up are constructed of lumber. There are three prominent forms of wood frame construction used today:

- **Balloon frame.** Studs run two or more stories high from the foundation to the eave lines. The floor assemblies are tied to the stunts by resting on a board laid horizontally called a ribbon. Buildings constructed in this manner are very susceptible to fire spread if it enters the wall or floor assemblies.

- **Platform frame.** Platform frame construction is first built on a platform generally of 2×12-in. boards. The platform contains the floor joists of the first floor. The wall assemblies for the first floor are then built on that platform. The specific wall assembly only goes as high as the ceiling of that particular floor, usually 8 to 10 ft. Another platform containing floor joists for either the attic or for the second floor is built on that wall assembly. Then wall assemblies are built on that platform, and so on.

- **Truss frame.** Truss frame is engineered construction in which the roof, floor, assemblies, and studs are tied into a unitized frame. The studs are an integral part of both the roof and floor assemblies. Truss frame is also called lightweight wood construction because the pieces of wood used are smaller than the typical 2×4-in. lumber used in platform or balloon frame construction.

Of the three forms of wood frame construction, the majority of homes existing today are platform construction. Many older homes in your jurisdiction may balloon, and very new homes may be truss frame construction. Combinations of platform and truss are certainly in existence today. They have lightweight truss roof assemblies as well as lightweight engineered joists.

The next most abundant materials used for home construction are brick or CMU. Don't get confused with a brick veneer. They use wood-framed balloon or platform-type construction with an exterior weatherproof coating of brick. This type is easily identifiable by the absence of header courses in the brick veneer.

Steel is a less frequent material, but you'll still see it used in some single-family residential occupancies. To the best of my recollection, I've never seen a single-family residential occupancy where the load-bearing walls were constructed of steel. I am sure that there are some across the United States, but I do not ever remember coming across one.

There is a new trend in building construction called a *fortified home*. They are constructed to withstand natural disasters that occur in some specific areas of the United States. In the Midwest, they are designed to withstand tornadoes. In the Southeast, they are designed to resist hurricanes. Some have poured concrete exterior load-bearing walls. If it is very difficult to tell these homes apart once they have been constructed; they appear to be typical homes. Many homes on some coastal areas have hurricane windows that are extremely difficult to vent (break) by conventional means. The only way to know for sure is to watch buildings go up in your community. When you find one that appears to be fortified, stop and ask the builders questions.

From the above we can make the following assumptions related to the building construction in single-family residential occupancies:

- The majority are wood frame.

- Most of those are platform-constructed homes built between 1940 and the present. They have inherent fire stops that resist the spread of fire from one floor to another.

- Homes in your community built prior to 1940 are likely balloon frame construction. These are very susceptible to fire spread in all directions if fire enters the wall or floor assemblies.

- Homes built over the last 20 years may have truss assemblies. Most new homes built today have truss roof assemblies. Many have engineered beams for floor assemblies. Some are true truss frame constructed buildings. This type of construction may or may not lead to rapid fire spread, but certainly are more prone to early collapse due to the size of wood used in the assemblies.

Prioritizing Search in Single–Family Residential Occupancies

Single-family residential occupancies are the place where most civilians die due to fire in the United States today. From that standpoint, searching these occupancies should take a very high priority at working fires in these occupancies. There are several things to consider. Just as too many civilians die every year in single-family residential occupancies, too many firefighters died searching these occupancies for victims that were actually not there. Many others were seriously injured searching for obviously deceased victims. Most of the reports that I see of injured or killed firefighters prove to be firefighter fatality fires only. The report usually indicates that firefighters *believed* there were civilians inside.

Here are some cues for prioritizing search in single-family residential occupancies:

- Does the house appear to be vacant or occupied? In my vocabulary, a *vacant* building is one that no one occupies at the time. In a vacant house there is minimal if any furniture. The heat and power are usually shut off and the building should be locked and secure.

- An *unoccupied* building houses people, there are furnishings present, and the heat and power are on and working. However, no one is inside at the present time. In an occupied building or home, people are inside the building at the present time.

- Just because a structure is vacant, it does not mean that there aren't human beings inside the house at the time of the fire. The homeless often seek shelter in vacant houses. Unless you see or hear a victim inside an obviously vacant home, treat it as a vacant building. I want to repeat that! Unless you *see or hear* a victim inside an obviously vacant home, treat it as a vacant building. Too many firefighters are killed and or injured searching vacant buildings because civilians say "they think (or even know) someone is in the building." In these instances, unless you see or hear a victim inside an obviously vacant home, treat it as a vacant building. As such, search can be a much lower priority. If we could accurately document the number of saves that were made in vacant buildings compared to the number of firefighter deaths and injuries that occur, the results will show it's not worth the risk.

- I prioritize search more highly on my to-do list at night time and early morning fires than I do at daytime fires. If there appear to be children living in the house, search is always a higher priority, regardless of the time of the fire. Kids tend to take naps throughout the day, so you never know when they may have lowered awareness of their surroundings.

- Toys, bicycles, and swing sets tend to indicate that children maybe living in the house. Handicapped accessible ramps may indicate that people with disabilities or the elderly reside in the house. I also hope that the first firefighter in the door (probably assigned to attack) notifies the incident commander if he or she notices cribs, playpens, walkers, or wheelchairs inside the house.

These cues are not all-inclusive, but they are the leading indicators that I use to prioritize a search in a single-family residential home.

Searching an Older Traditional Two–Story Home

This is the true bread and butter fire for many departments in the United States. Figure 18–1 shows the typical two-story, three-bedroom house with one bathroom located on division two. The first floor contains a living room, kitchen, half bath, and a dining room.

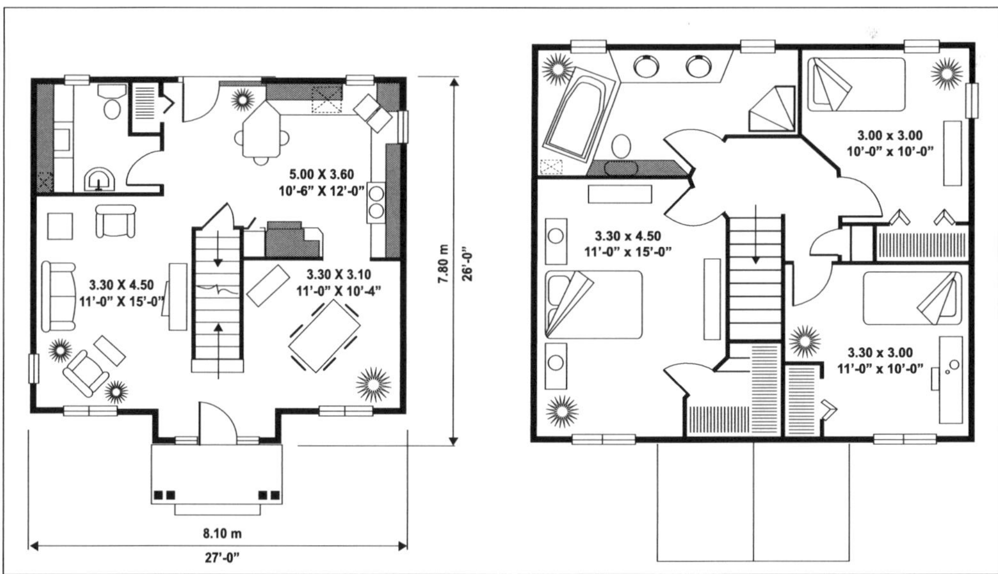

Fig. 18–1. Floor plan for a two-story, three-bedroom house

Standard search

Searching division two. The crew enters the house following the attack line, most likely through the front door. If there is no fire on division two, locate the stairway and advance toward the second floor hallway. If the crew is a four-person crew, we have two options.

In one option, the crew can split into two teams of two, all searching the same room. Two firefighters search in one direction and the officer and the other firefighter search in the other direction. The officer remains near their wall and the other firefighter moves laterally toward the center of the room. They perform a predetermined number of side crawls, then return to the officer.

The other option is that the crew remains intact and the officer maintains contact with his or her wall. The rest of the crew is positioned laterally toward the center of the room. Then they move around the room in a line.

If using a three-person crew, search one room at a time. Send two members to the left and one to the right, or vice versa. Alternatively, they can work together doing a left- or right-handed search while the other members work out laterally toward the center of the room.

Searching division one. Enter the house following the attack line, probably through the front door. The crew can either split into two teams or stick together as a unit. Go to the area closest to the fire where savable victims could be and begin your search. Cover that room or area in its entirety before moving on to the next one. The officer decides which room comes next based on the search plan.

If using a three-person crew, search together as one team doing a left- or right-handed search throughout the entire first floor. The officer has two options. One is to search each room completely and then move onto the next room. In performing this option, two firefighters go in one direction and the other goes in the opposite. They all sweep out toward the center of the room until the meet somewhere at the opposite side of the room.

With the other method, they can cover the entire first floor starting using a left- or right-handed search until the floor has been covered. In performing this evolution, the crew must stay together. The officer must stay on his or her wall and the other firefighters position themselves laterally toward the center of the room. In my opinion, the first option is safer and assures all rooms are covered.

Team search

Due to the size of the house and its layout, I would not expect a crew to use team search in this type of occupancy. You are never more than 13 ft. from a door, and you're usually much closer to the previous window. Notice I said previous window. You should almost always return to your last known window or door if you need emergency egress. Unless you can see an exit ahead of you, this is the surest way to get out of a dangerous situation.

The process of team search is too time consuming. The intent is to maintain orientation with the point of egress, which is generally not too difficult in a home. Additionally, there are many obstructions and hindrances to tangle and snag on the search rope

VES

In the community where vent/enter/search was developed, more than 80% of the homes were single-family, two-story residential occupancies. The original VES method used one firefighter. As common sense and standard fire department practices tell us, very little should be done on the fireground with only one firefighter. If you believe victims are located in bedrooms on division two, then two firefighters can use the oriented VES to rapidly search the sleeping areas.

Two firefighters ladder division two off of the front porch. The first firefighter up the ladder clears the entire window and sounds the floor. After that, the searcher enters the bedroom and immediately closes the bedroom to the hallway if it is not already closed. The second firefighter climbs the ladder and takes a position on the porch roof at the window where the first firefighter entered. After reaching the window, the oriented firefighter communicates with the searcher to assure the searcher's well being.

After closing the door, the searcher searches the bedroom, beginning and ending at the hallway door. Once the search is complete, the searcher returns to the window and exits to the porch roof. If another bedroom is reachable from the porch, the searcher should enter that bedroom next and repeat VES.

If additional bedrooms are located in areas that are not accessible from the front porch roof, a ground ladder should be used to ensure that those rooms are searched as well. In order to accomplish this, the two firefighters ladder the bedroom window. The first firefighter up the ladder clears the entire window and verifies that the floor assembly is intact. The first firefighter then enters the bedroom and ensures that the door to the hallway is closed. The second firefighter climbs the ladder to the window and communicates with the searcher to assure his or her well-being.

After closing the door, the searcher searches the bedroom, beginning and ending at the hallway door. Once the search is complete, the firefighter returns to the window and exits onto the ground ladder. At this point both firefighters should descend the ladder and repeat this process on any other bedrooms of the house.

Vent/enter/search is designed specifically for sleeping areas. Under most circumstances, it should not be conducted in areas other than bedrooms.

Oriented search

Searching division two. To start a search using the oriented method on division two, the crew should enter the house following the attack line through the front door. The oriented officer leads the crew up to the second-floor hallway. The oriented officer instructs a firefighter to start a search in the AB quadrant of division two, which in this case is the master bedroom. The searcher informs the oriented officer if the search is left- or right-handed. In this instance, the searcher tells the oriented officer that it is a four-walled room, then begins a one-person search

The oriented officer locates the next room to be searched. Once it has been identified, the oriented officer leads another searcher to that doorway. In figure 18–1, it is the division two bathroom. The searcher indicates the direction of search to the officer and that the room appears to have four walls. The searcher should inform the officer that the room is a bathroom as soon as it is realized. This tells the oriented officer that it will not take long to complete this room. If this is a four-person crew, the oriented officer should take the last searcher to the bedroom and the CD corner of division two. Again, the searcher informs the oriented officer of the search direction and the number of walls in the room.

It may be harder to determine the number of walls in this room. If the searcher runs his or her arms down the walls on both sides of the door, it should be apparent that something isn't quite right with the room. The wall on the right appears to go off at a 90° angle, and the wall on the left goes back in a different direction. At a minimum the searcher should tell the oriented officer this is not a square room before proceeding the search.

At this time the oriented officer needs to go back and check the progress of the first and second searchers. If time permits, the oriented officer should briefly survey the atmosphere for excessive heat. I would not expect the oriented officer to stand in the hallway, but rather just raise an arm and expose a portion of the skin only long enough to determine the heat in the area, then immediately cover the exposed area again. Either that, or the officer should pull his or her hood back and rapidly expose some facial skin to test the environment.

The searcher responsible for covering the bathroom should be finished relatively quickly. Bathrooms are not as difficult to search because they are small and don't contain moveable furniture or obstructions. When searching a bathroom, always remember to sweep the bathtub quickly with a hand to ensure that no one is hiding there.

As soon as the oriented officer gets a searcher who has completed the assigned room, the oriented officer should lead the searcher down the hallway in a clockwise direction to the last bedroom. In figure 18–1, that's on DA corner of division two.

This searcher indicates the number of walls, which is four. The searcher then begins search. At this point, the oriented officer returns to the other two searchers to check on their progress. As they finish their respective searches, the oriented officer should keep them in the hallway close by unless the officer feels it would be productive to send them to the last bedroom to aid in search.

Once all rooms have been covered on division two, the oriented officer would either lead the crew up to the attic or take them down to division one. As a reminder, my guideline for attic search is to see if there is a permanently affixed stairway leading to the attic. After a quick initial sweep at the top of the attic stairs it should be easy to determine if there could be people up there. The appropriate benchmark should then be given to command, "Search to command, all clear on division two, going to division one."

Searching division one. The oriented officer used the stairway and the hallway of division two as the point of orientation to the means of egress from the building. That means always knowing where you are in relationship to the stairway and where the stairway is in relation to the door out of the building. Because there is usually no hallway in most first floors of single-family residential occupancies, the oriented officer must choose another fixed object with which to maintain orientation to the way out. In most cases, this will be the stairway assembly to division two.

If you look at figure 18–2, you'll notice that the walls that frame in the stairway going up to division two are located in the center of division one. If you follow these walls around in a clockwise direction, you'll come back to the same starting point. Realizing where the base of the stairway is in relation to your exit provides orientation to the path of egress.

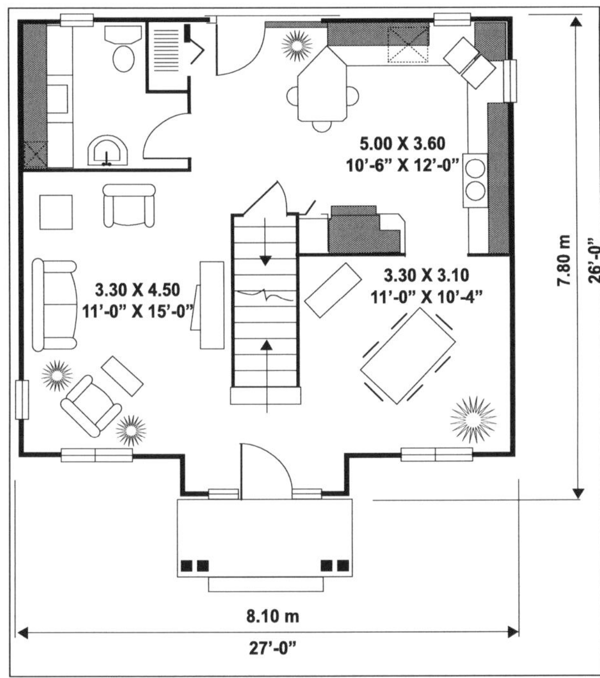

Fig. 18–2. Division one of the two-story single-family occupancy

To search division one, the oriented officer needs to locate the base of the stairwell. One searcher needs to be sent to the left to the search the room at the AB quadrant. Then have a one firefighter conduct a search in the room immediately to the right or in the AD quadrant of division one. The member conducts search again, first telling the oriented officer the direction of search and the number of walls, if possible. As you can see, determining the number of walls in this room may be difficult.

The oriented officer then leads the remainder of the crew in a clockwise direction toward the back of the stairwell wall system. You may use a counterclockwise path if it works best, depending on the structure and situation. The officer instructs a searcher to search the BC quadrant which, in this case, is a small half bath. If I were the oriented officer at this time, I communicate to the searcher in the AB corner to determine progress, then lead my last searcher to the kitchen area to start a search. If time permits, the oriented officer should check for environmental safety concerns, including heat in the upper atmosphere and the floor.

When a searcher is available, the officer should again move clockwise around the stairwell framework to the last area to be searched on division one. The officer should give the appropriate benchmark including where they intend to search next, if applicable.

When the searchers are completed, the officer should search on division two (if not already completed due to the time of the fire and possible location of victims). If division two has been searched, then move back to the interior basement stairs. In this case, it is at the opposite end of the stairwell framework in the kitchen area. The search will be complete when they've searched the basement area.

Searching a Split-Level Home

A split-level house is defined as one in which some rooms are half a floor above or below the others (fig. 18–3). The access to all floors is generally in the middle of the building. The lower level commonly consists of a garage with a stairway that leads up to the middle level. The level above the garage usually contains the living room and a kitchen. The level above that contains the sleeping areas (fig. 18–4). These are relatively easy houses to search.

Fig. 18–3. A split-level home exterior

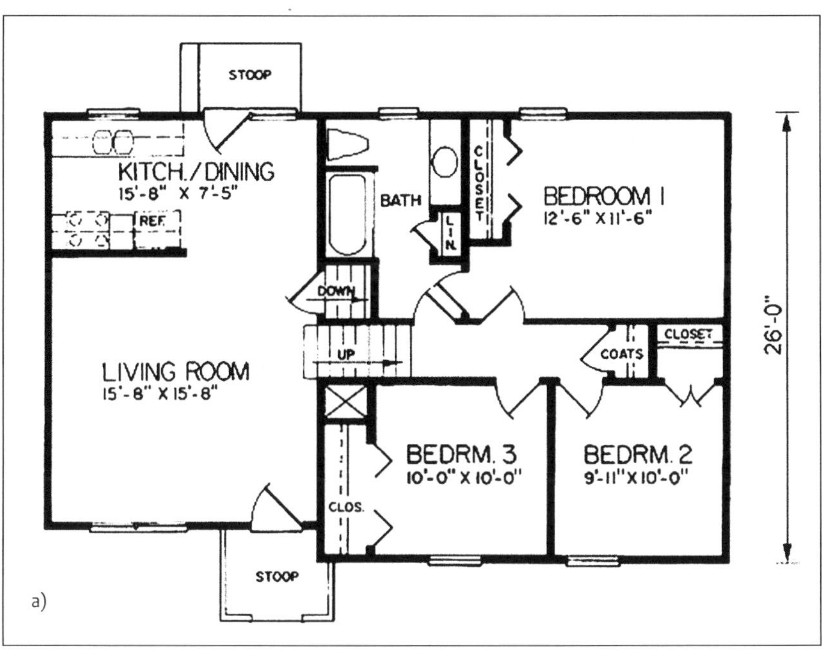

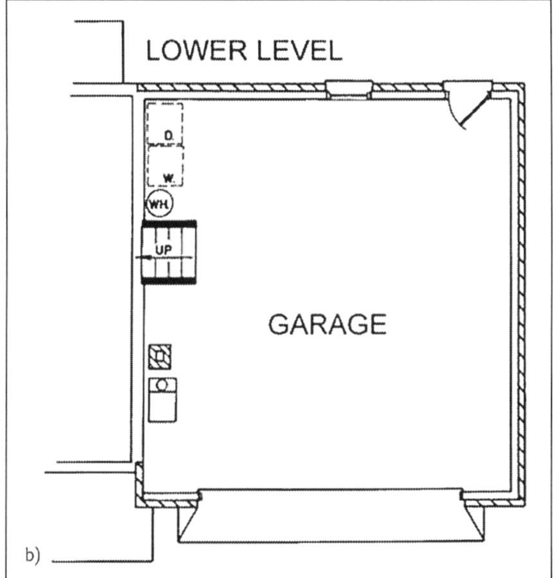

Fig. 18–4. Floor plan of a split-level home: a) main and upper level b) lower level

One concern when fighting fires in split-level houses is how we verbally label each division or level of the house for communication purposes. I realize in conforming with NIMS we should designate them as division one, two, and three, with division one being the basement, division two being the living room/kitchen level, and division three being the bedroom areas. I am sure in some jurisdictions they simply refer to it as the garage level, the living level, and the sleeping level. I have no problem with this. However, I am sure that some of the ICS purists disagree. What matters is that everyone understands how you differentiate one floor level from another. To keep this textbook as uniform as possible, I refer to them as division one, division two, and division three. As you may have noticed in previous chapters I refer to the attic and basement using their common name. I don't have a problem with this. In fact I believe it makes the fireground less complicated. If you have a problem with it, don't do it!

Standard search

Searching division three. In this scenario we assume search's intent is to check the sleeping areas first. Because of the layout of most split-level homes, the officer knows that the stairway leading to the upper half or division three (the sleeping area) is usually located straight ahead of the front door. In this case it is along the wall located at the door entrance. The officer leads the crew up to the sleeping area and, assuming a four-person crew, they have two options.

The first is for the entire crew to enter the first room encountered. Two firefighters go to the left and two go to the right. They meet somewhere on the opposite wall and then leave the room, moving to the next encountered room. In the other option, they conduct a search moving in a single direction with the officer moving along their wall and the remainder of the crew fanning out toward the center of the room.

If using only a three-person crew, search the sleeping areas having two firefighters go in one direction and the officer in the other. They all meet somewhere in the middle of the room and then head back out to the next area to be searched. It is imperative that they remain oriented with that interior division three hallway. That always indicates the direction and location of their exit.

When division three is completed, the officer should give an "all clear" on division three to the incident commander. Then the crew moves to division two in order to begin a search there.

Searching division two. In understanding the layout of split-level homes, the officer may opt to split the four-person crew into two teams. One team starts to the right at the bottom of the stairway heading toward the kitchen area, and the other team heads to the left to search the living area. If the officer sends two firefighters to the right toward the kitchen area, they should be instructed to continue on with search in the kitchen when they reach it. After that, exit the kitchen and start search in the living area. They meet back with the officer somewhere in the living area when they're done.

Another method would be for the crew to remain intact. The officer stays on his or her wall and the remainder of the team fans out toward the center of the room. They proceed around the room as a line in unison.

If using a three-person crew, the only option is to have them search together as a team, either going to the left or to the right. This search is conducted with the officer following the wall in order to maintain orientation while the two firefighters fan out laterally toward the center of the room. They sweep the area, then move back toward the officer's wall. At this time the search officer should again provide the incident commander with an "all clear" prior to moving to another division.

Searching division one. Once the crew has covered division two, they should move down to division one to see if there is a living area adjacent to the garage. In this instance there is not, but some homes have a family room or other living area there. They can quickly cover the garage and then exit the building using the front door unless there is an excellent visibility in the basement and the garage door is already open.

Team search

Due to the size of the house and its layout, I do not expect a crew to conduct the search using the team search method in this occupancy.

VES

Most split-level homes have no porch roof leading to sleeping areas. Generally the sleeping areas in split-level homes are located at the highest portion of the house. Individual bedrooms have one or more windows that need to be laddered in order to be accessed.

Two firefighters ladder the bedroom window. The first firefighter up the ladder clears the entire window, sounds the floor, and enters the bedroom. The searcher must immediately locate the door separating the bedroom from the hallway to ensure that it is closed. The second firefighter climbs the ladder to the window and communicates with the searcher to assure their well being.

After closing the door the searcher searches the bedroom beginning and ending at the hallway door. Once the search is complete, the searcher should return to the window and exit onto the ground ladder. At this point both firefighters should descend the ladder and repeat this process at any other bedrooms of the house.

Oriented search

Searching division three. The oriented officer leads the crew up to division three, most likely by entering the front door following the attack line. Creating a plan prior to entering, the officer has envisioned the typical split-level home and makes an educated guess that the stairwell is located somewhere in the center of the home ahead of the front door.

Depending upon the size of the crew, the officer sends the first searcher to the first door encountered on division three. In this instance, it is a bathroom. The searcher informs the officer of the intended direction of search and the number of walls in the room. In this instance, if the searcher goes to the left it would seem like a four-walled room. They should also inform the officer that it's a bathroom, indicating they will be done in a relatively short amount of time.

Once the searcher begins, the oriented officer locates the next room to be searched and sends a firefighter into that room. If the first searcher has not completed the bathroom, the oriented officer should communicate with the first searcher to ensure the searcher's well-being, then locate the next room to be searched. If there is another searcher because it is a four-person crew, the third searcher is sent into the next room encountered off the hallway. In this instance, it is the front bedroom on the AD corner of the house.

While waiting for a searcher to become available, the oriented officer should take this time to check heat conditions mid-level in the hallway and the amount of heat on the floor. Once a searcher becomes available, the oriented officer leads them to the last room on division three to be searched. Once that last room is completed, the officer gives the appropriate benchmark and begins searching on division two.

Searching division two. To search division two, the oriented officer should position the crew at the bottom of the stairway leading from division three to division two. This allows them to maintain orientation with the path of egress which, in this instance, is along the left wall to the door.

If using a three-person crew, the officer sends one firefighter to the left and one to the right with the intent of covering the living room and kitchen areas. Because of the kitchen size, the officer should inform the searcher who went right to continue on into the living room when finished. The two searchers should meet somewhere in the living room.

If a four-person crew (including officer) is used, the officer can send one searcher to the right to start a search in the kitchen area. Then, depending up the wall layout, send two firefighters to the left to conduct more of a standard search with one firefighter maintaining contact with the wall on the left and the other firefighter working out laterally toward the center of the room and then back to the wall. It doesn't take long to provide brief instruction to searchers in order to ensure that they conduct the most appropriate search. When division two is complete, they work their way down to division one and conduct a search in a similar fashion. The oriented officer maintains position at the bottom of the stairway. This ensures that orientation with the path of egress is maintained.

Searching a Ranch Home

Ranch homes are homes with one story above grade that contains both living and sleeping areas. Ranch homes may or may not have a basement, depending upon several factors. Ranch homes can be built on a slab, with the crawl space or a full or partial basement. A home with the living and sleeping areas all on one floor may have a full basement. This could appear to be a two-story home from the rear, but it is still considered a ranch home.

Ranch homes can be quite difficult to search because of their size and the remote location of some areas in relation to the door of entrance. It is generally very easy to differentiate the living areas from the sleeping areas in a ranch home. However, unlike a split-level or traditional two-story, it is hard to estimate the floor plan of many of these homes. Because of this, I cite several examples in this section and give a detailed explanation on one, with more brief explanations of how to conduct search on the others.

Figure 18–5 shows a typical 2,200-sq.-ft. ranch home that I consider of medium size. This plan is a relatively simple ranch to search. As with ranches of this size, the bedrooms are located in one area, usually at the opposite end of the house from the garage and living area (fig. 18–6). In these homes, locate the garage to simplify the search.

Fig. 18–5. Exterior of a ranch home

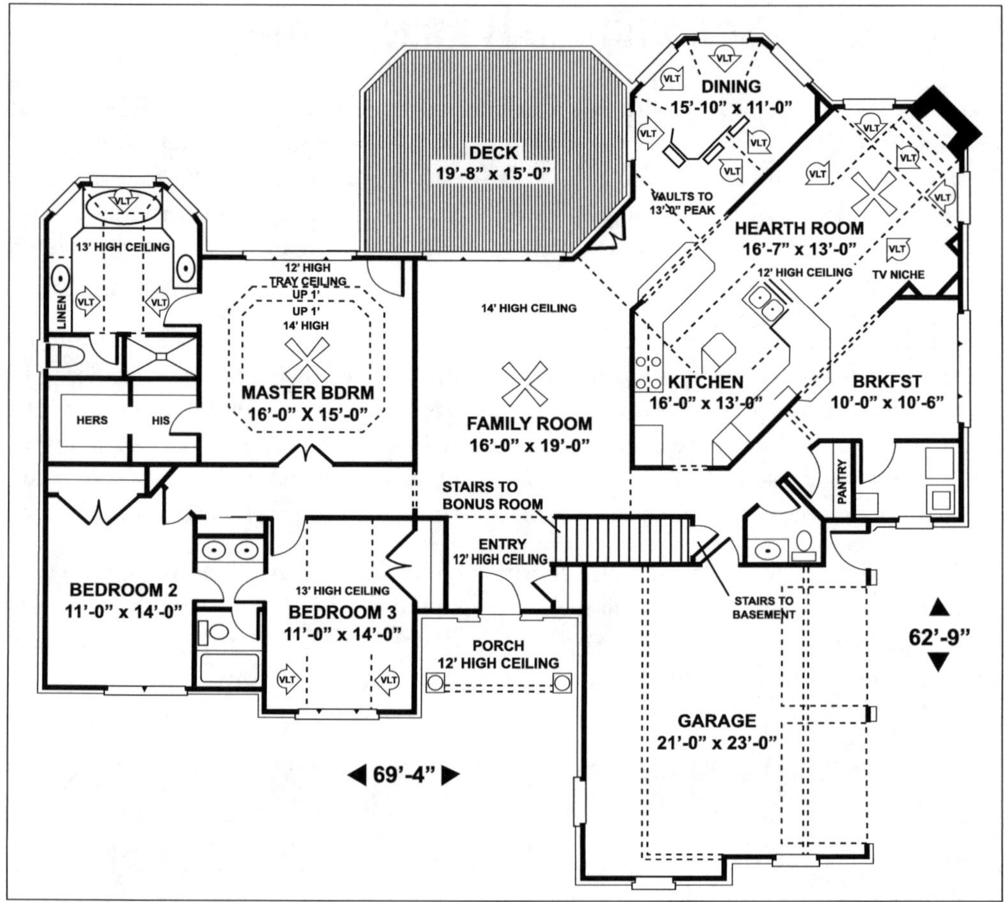

Fig. 18–6. Floor plan of a ranch home

Standard search

Assume that the fire occurs at a time when it is best to search the sleeping areas first. The crew enters following the hoseline in through the front door. Since the garage is on the AD side of the house, the sleeping areas will be on the AB and BC sides. If the officer feels more comfortable pulling an 1¾-in. line in to act as a point of orientation, so be it. The line should be pulled and stretched dry until the crew locates the hallway off of which the sleeping areas are located. At that point the officer can opt to have the line charged or leave it dry in a location near the entrance to the hallway.

Once the hallway is located, the officer takes the crew down to the first bedroom. Using the wall on the left for orientation, they enter the bedroom on the left side of the hallway first. If the crew is a four-person crew, two members go to the left and the officer and the other member go to the right, meeting somewhere on the opposite wall. Once the room is complete, they exit together and locate the next room. In this case, it would be the next bedroom. They should continue until the sleeping area is complete.

If using a three-person crew, they can enter the first room encountered, and one firefighter heads in one direction while the officer and another member searches in the other direction. They meet somewhere in the middle and all exit the room together. This pattern is continued until the entire sleeping area is completed and the search crew is back at the hallway entrance.

Understanding and envisioning the typical layout of this type of ranch house, the officer believes the next area to be searched is the living areas. This includes the living room, great room, and the formal dining area. In conducting the standard search, the officer should be participating in order to visualize the items that may be encountered to ensure that the house is laid out as envisioned. In this case, that could be sofas, chairs, coffee tables, and a long table and chairs in the dining area.

To search this area, the officer has two general options. The search can start from the hallway that leads to the sleeping areas and the officer can move in a clockwise or counterclockwise direction until the living area and dining room have been covered. The other option is to follow the wall to locate the hallway wall near the front door, then conduct a left- or right-handed search to cover the living and dining room areas.

When the crew encounters the kitchen, the officer can opt to search it along with the living and dining area, or they can continue to cover the living and dining area exclusively and return to the kitchen area as soon as the other areas are complete. I believe that, due to the size of the kitchen and breakfast area, they should be considered separate areas.

Once the kitchen, breakfast, and the utility areas leading to the garage have been covered, the garage should be searched last. Upon completing the garage search, the crew can exit the garage if visibility permits, or they can work their way back through the kitchen to the living area and out the door they entered.

If the officer opted at the beginning of the search to take a hoseline along as a point of orientation, that line can be taken to the entrance to the hallway leading to the sleeping areas. It can either be left there or taken back to the door, depending on where the officer chooses to start search of the living and dining area. It can also be taken to the kitchen area to remain as a point of orientation to the exit while the crew searches the kitchen and garage. When completing each individual area of the search, the officer should give a benchmark of "all clear" in to the incident commander.

Team search

A version of team search could be done in a house of this style. A search rope is used in the same fashion as the 1¾-in. hoseline was in the above scenario. The rope is anchored to some fixed point outside of the house. The rope should only be played out of the bag enough to reach from the hallway to the sleeping areas initially. I would leave the anchor at the bag and then conduct a standard search of the sleeping areas using the hallway as my point of orientation. Do not play the rope out while conducting the search in the sleeping areas, as it takes too much time and effort away from conducting the actual search.

When the search of the sleeping areas is complete, leave the bag and anchor in its current location at the entrance to the sleeping areas. Then conduct a left- or right-handed search from that point. Alternatively, take the search bag back to the point of entrance to the house and leave it and the anchor there, then conduct a left- or right-handed search back to the entrance and the anchor.

VES

In chapter 8 I discussed that vent/enter/search was being used successfully and safely in Estero, Florida, using a combination of the original method of VES and the oriented method of search. A majority of homes in that area are ranch style homes. When using VES, the two firefighters employ an attic ladder or other means in order to gain access to the bedrooms.

Two firefighters ladder the bedroom window with an attic ladder. The first firefighter up the ladder clears the entire window and then sounds the floor. After that, the searcher enters the bedroom and immediately locates the door separating the bedroom from the hallway to assure that it is closed. The second firefighter should remain at the window and communicate with the searcher to assure the searcher's well being.

After closing the door, the searcher checks the bedroom, beginning and ending at the door hallway door. Once the search is complete, the searcher should return to the window and exit onto the ground ladder. At this point, they should repeat this process at any other bedrooms in the house.

Oriented search

Assuming that we start search in the sleeping area, the oriented officer leads the crew into the house through the front door, following the hoseline. They then work their way toward the sleeping area. By properly reading the building for search and creating a plan prior to entering, the officer realizes that the sleeping areas are on the B side of the house.

The oriented officer finds the hallway leading to the bedrooms and uses that as the point of orientation to the path of egress. When the oriented officer comes to the first room to be searched, one firefighter is sent in to conduct search. The searcher should inform the officer of the direction of search and that it is a four-walled room. The oriented officer should acknowledge the communication, then moves on to the next room where repeats the practice with the next searcher.

If this is only a three-person crew, then the oriented officer should go back to the first room being searched and check on the progress of the searcher. If using a four-person crew, the oriented officer leads the last searcher to the next room to be searched. In this case it's the master bedroom. The officer should also check for heat and smoke conditions by quickly exposing a portion of skin, both above the head and on the floor.

After completing the assigned room, the searcher should enter the hallway and notify the oriented officer that search is complete. The oriented officer instructs the firefighter to crawl forward past two doors to his or her location. This will take the member past the linen closet and second bedroom to the doorway of the master bedroom. The oriented officer then sends the first firefighter into the next room to begin another search. When another searcher is available, the oriented officer instructs them to begin search in the opposite of the first searcher. As communicating is very important in this method of search, the oriented officer should tell the first searcher in the room that another searcher will be sent in to start a search in the other direction.

When the sleeping area has been covered, the oriented officer moves back to the hallway leading from the sleeping area to the living room and, as in the standard search, there are two options available to search the living area. They can start a search maintaining orientation with the exit utilizing the hallway and its orientation to the front door, or they can move back to the front door and begin search from that location.

The officer realizes that a typical ranch house has a living area that probably contains a family room and a formal dining room. As long as the oriented officer can maintain voice contact with the searchers, the officer can remain in this position while they search. The officer could request that the searcher working to the right announce when the kitchen is located so the officer knows where the kitchen is in relation to the door.

When the searchers complete this area, the officer leads them to the kitchen area to search. To get there, the officer follows a wall along the dining room and up to the entrance to the kitchen area. The officer still has an orientation point with the doorway and exit. As the searchers check the kitchen and breakfast area, the oriented officer can locate the door to the garage and also take the time to again check heat and smoke conditions.

Once the kitchen area is covered, the oriented officer can send the searchers into the garage. The oriented officer maintains a position at the doorway between the kitchen and the garage. Upon completion of the garage the crew exits the building and the officer gives the final benchmark "all clear" to the incident commander.

Searching a Small Ranch

This ranch is set up similarly to the mid-size ranch we discussed above, with the exception of its size (figs. 18–7 and 18–8). This house is about 800 sq. ft. smaller. I would search this house with either the standard or oriented method in the same fashion described above. I do not believe I would use a form of team search using ropes or a hoseline in this house because of its size.

Fig. 18–7. Exterior of a small ranch house

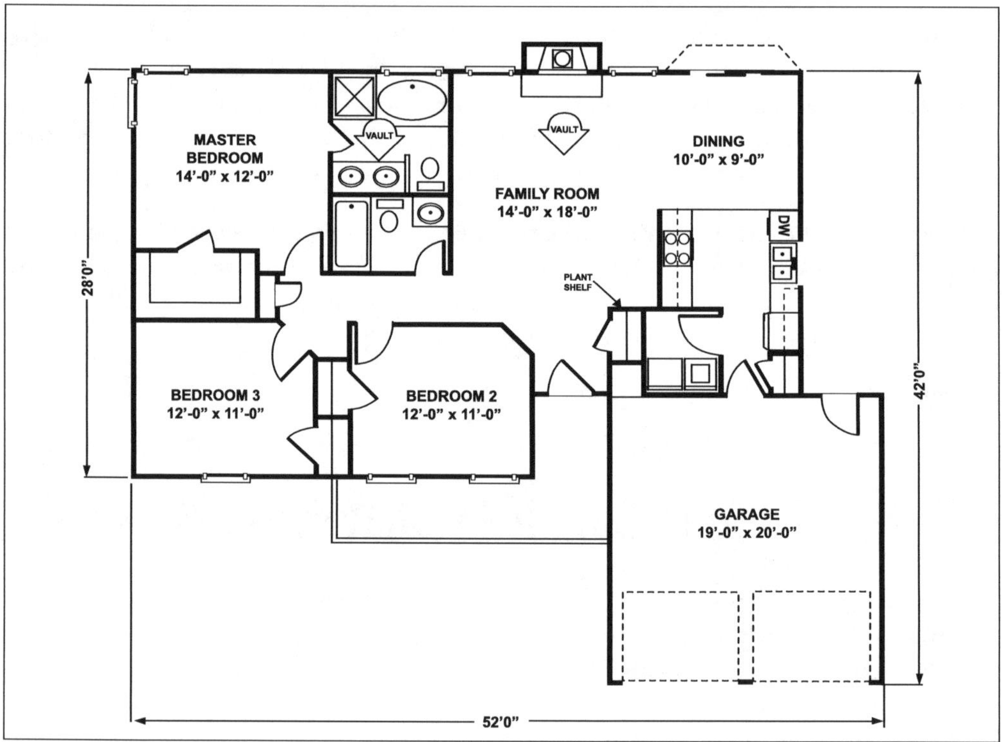

Fig. 18–8. Floor plan of a small ranch house

Searching a Large Ranch

This ranch house is a slightly under 5,000 sq. ft. (fig. 18–9). As you can see, it has an atypical layout. This house will be a challenge to search regardless of the method used. The master bedroom is on the C side of the house There is a bedroom behind the garage on the AD side, along with two other bedrooms on the AB side of the house. Entry to the house is behind the garage on the D side.

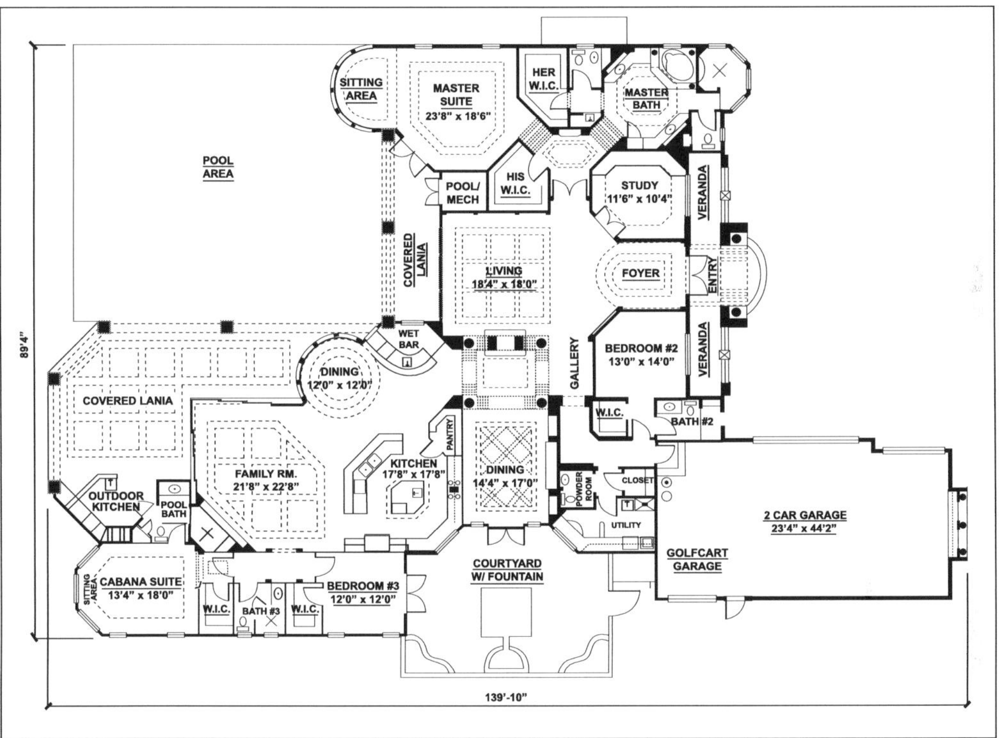

Fig. 18–9. Floor plan of a large ranch house

Before we even discuss the "hows" of this search, let's start by doing a little math. This house is 5,000 sq. ft., not including the garage area. In order to cover this house in our 15 minute maximum time frame, the crew has to search more than 300 sq. ft. every minute. That equates to searching a room slightly larger than 15×20 ft. every minute. That is a daunting task, especially in IDLH conditions. The master bedroom is more than 425 sq. ft., not including the closet and bath area.

My first recommendation is to have at least two crews to conduct this search if at all possible. Because of the size, I would probably shy away from a standard search. Instead I would keep the crew together searching one area at a time and use the oriented method of search. I would either use a rope or an uncharged hoseline as my point of orientation to the exit. In either case, the oriented officer should maintain contact with the rope or hoseline at all times. You must always remember that the tool used to maintain orientation with the exit is not used in the traditional sense. If you opt for a hoseline, keep it uncharged. This will ease the oriented officer's ability to maintain contact with it while the searchers operate. If I opt for a rope, I simply drag it with me a as I move forward through the house to use it as a guide to locate my exit.

In using two crews of two firefighters each, with the intent of locating and searching sleeping areas due to the time of the fire, I would instruct one crew to start to the left and the other to the right inside the main entrance to the house and follow hallways in an attempt to locate bedrooms. In this instance, the crew going to the left has to travel only a short distance through the foyer and gallery to find the first bedroom (bedroom #2) in this house. Once done there, the crew must travel past the dining room into the kitchen where the entrance to bedroom #3 is located. The crew going to the right has to travel through the foyer, past the study, and to the master bedroom area. While moving forward in search of sleeping areas, I would send a firefighter into any room encountered with the intent of identifying the type of room. A few sweeps with the hand should indicate the type. I expect a study to have more formal type chairs and a desk, as opposed to a bedroom which will have the standard bed, dresser, and end tables.

One final thought on searching large ranch style homes. The math just does not add up. This can be a very labor-intensive search that takes a lot of time. Many of you do not have the resources of a larger department. My answer to searching these occupancies in low staffing is quite simple:

> *Get a line on the fire and aggressively vent using PPV, then send crews in to actually "look" for victims. That's much more effective than blindly feeling for them. With 10 ft. of visibility, you can probably cover a 6,000 sq. ft. home in less than 10 minutes with one crew. Remember that every minute that fan runs, your visibility should improve, as does the condition of the air available to civilians.*

I realize that this may be a radical thought for today's fire service. As you'll see in future chapters, this will be a recurring theme. Short of substantially increasing staffing to conduct a search in these larger occupancies, I can't come up with any other methods that will be as effective. The key is to locate and darken the fire, then aggressively vent.

Searching New Two-story Homes with Bedrooms on Different Floors

A new trend in home design is single-family, two-story homes with the master bedroom on the first floor and the children's bedrooms on the second floor or in a walkout basement (figs. 18–10 and 18–11). Both should be searched the same way. We assume, due to the nighttime nature of the fire, that our intention is to search the sleeping areas first.

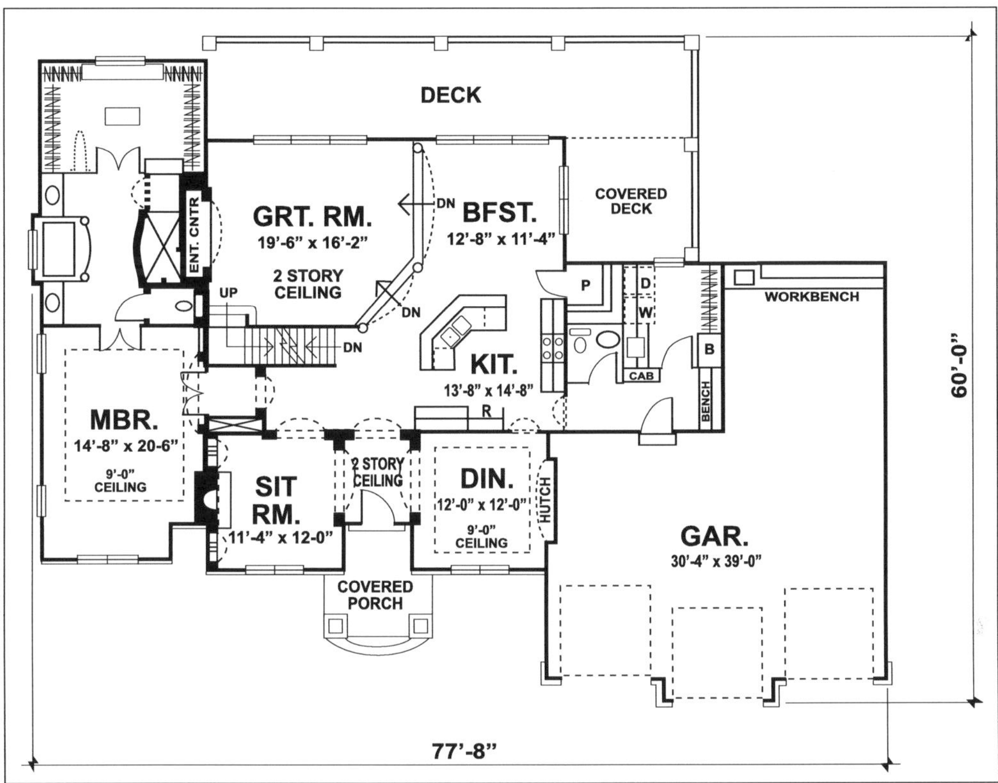

Fig. 18–10. The first floor of a two-story home with bedrooms on both floors

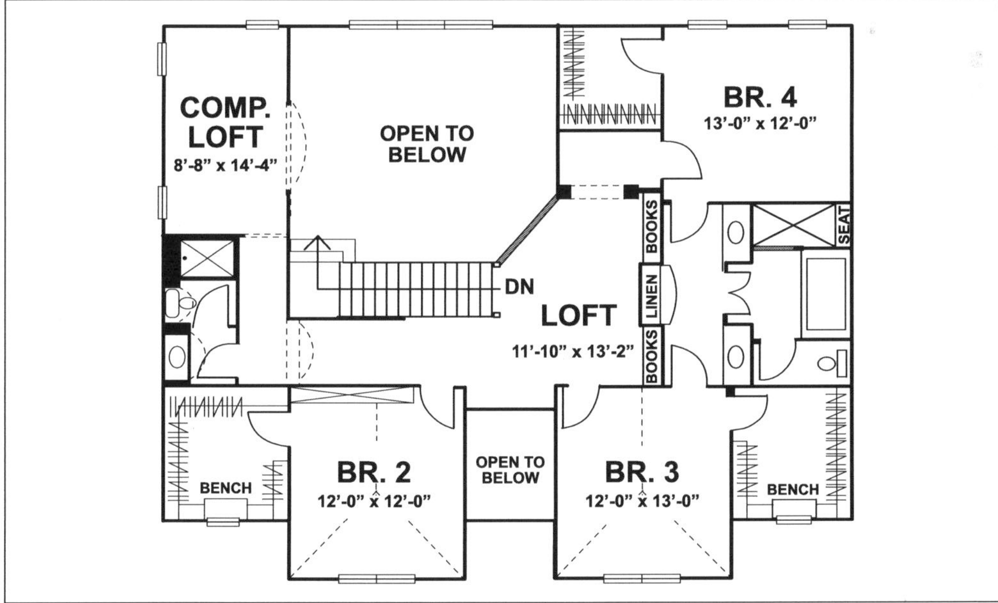

Fig. 18–11. The second floor of a two-story home with bedrooms on both floors

Standard search

If searching this house with the standard method of search, two crews of at least two firefighters should be used. One crew should head for division two and the other for the master bedroom on division one. The crew searching division two hits each bedroom, working off the hallway until division two is complete. The crew working division one attempts to locate the master bedroom on division one, bypassing all other rooms at this time. Upon reaching the master bedroom, the crew searches the bedroom conducting either a left- or right-handed search. Upon completing the sleeping areas, both crews should search the remainder of division one in a logical, appropriate manner.

If staffing does not allow for two crews to be employed, or if the crew assigned is not large enough to split into two teams of at least two firefighters, my gut tells me to search for children on division two first. I realize this house could be occupied by a couple, or even a single person with no children. If that's the case, they will probably be on division one in the master bedroom. Perhaps some cues from the outside, such as a lack of toys in the yard or porch may sway my decision.

Team search

I don't see a practical application for team search in this house. Division two is a very simple area to conduct a search. Maintaining orientation with the stairwell and hallway on division two allows for easy egress from the building. Division one is a little tougher, but if you maintain orientation with the wall that surrounds the stairwell, you should be able to direct a searcher into every portion of division one while maintaining voice contact.

VES

When searching these newer two-story homes with bedrooms on multiple floors, VES can again become an effective tool. A combination of extension ladders and attic ladders may be needed in order to facilitate VES. Depending upon the location of the sleeping areas, one or more of the methods described above should be used if VES becomes your evolution of choice for searching this type of house.

If the house has bedrooms off a porch roof, two firefighters would ladder the porch roof and climb onto the roof and up to the windows. The first firefighter up the ladder needs to clear the entire window. After ensuring that the division two floor assembly under the window is intact, enter the bedroom and immediately locate the door separating the bedroom from the hallway to assure that it is closed. The second firefighter should be positioned on the porch roof at the window where the first firefighter entered. Once positioned outside the window, the second firefighter communicates with the searcher to assure their well being.

After closing the door, the searcher should search the bedroom beginning and ending at the door separating the bedroom from the hallway with either a left- or right-handed search. Once the search is complete, the searcher should return to the window and exit onto the porch roof. If another bedroom is available from the front porch roof, then the searcher should enter that next bedroom and search as described above.

If the house has bedrooms on division two but no porch roofs, two firefighters ladder the bedroom window using an appropriately sized extension ladder. The first firefighter climbs up the ladder, clears the entire window, and sounds the floor assembly. After ensuring that the floor assembly under the window is intact, enter the bedroom and immediately locate the door separating the bedroom from the hallway to assure that it is closed. The second firefighter should climb the ladder to the window and communicate to verify the searcher's well-being.

After closing the hallway door, the searcher clears the bedroom beginning and ending at the hallway door. Once the search is complete, the searcher should return to the window and exit onto the ground ladder. At this point, both firefighters descend the ladder and repeat this process at any other bedrooms of the house.

With bedrooms on the first floor, two firefighters ladder the bedroom window, using an attic ladder or by other means. The first firefighter clears the entire window and verifies that the floor assembly under the window is intact. Then the searcher should enter the bedroom and immediately locate the door separating the bedroom from the hallway to assure that it is closed. The second firefighter should climb the ladder to the window and communicate to assure the searcher's well-being.

After closing the door, the searcher checks the bedroom using a left- or right-handed search, beginning and ending at the door to the hallway. Once the search is complete, the firefighter should return to the window and exit onto the ground ladder. At this point, both firefighters should descend the ladder and repeat this process at any other first-floor bedrooms in the house.

Oriented search

Two crews will be available to search this home in an ideal situation. Absent that, the oriented officer needs to decide whether to search division two or division one first. If I were the oriented officer and believed that children were living in the home, I would opt to search division two first.

Search using the oriented method as I described above. The key to determining how to search these larger single-family houses where sleeping areas are remote from one another is to make educated guesses about what type of individuals are in the house. When you get down to it, protecting children and young adults has always been a priority for the fire service, while always attempting to do the best for the most.

Maintaining Continuity of Search

It all seems easy until you find a victim. Although that is our goal, when we do find someone it seems as if a huge obstacle is then placed in front of us. If we can quickly remove the located victims and return to the exact point where we found that victim to continue our search, then we are indeed doing the best for the most. If, however, we get caught in the trap of locating and removing a victim, but don't continue from where we left off, we jeopardize the lives of any yet-to-be-located victims.

This is where the job of the oriented officer proves itself. The oriented officer is responsible for knowing where we have searched, where we are currently searching, and where we plan to search next. If we find someone, we must ensure that we aren't missing any areas when returning to the search. The oriented officer is responsible for keeping track of the progress so a searcher can pick up where the search left off, whether it's the same firefighter or a different one.

That information can easily be identified with one quick question as the searcher drags out the victim. "What wall were you on?" If the searcher can rapidly reply, "Right-hand search on wall three," then the oriented officer can direct whoever returns to continue the search to start a right-hand search beginning on wall three. That way we can avoid duplication of efforts, not miss anything, and save time.

There are several key factors in maintaining continuity of search when a victim is located:

- Identify the searcher who locates the victim and where the searcher was in the room when the victim was found.

- Ensure that the appropriate number of searchers is available to remove the victim to the outside or a safe haven.

- Never leave a single firefighter inside a building alone. (I realize I said *never*. In this case it works.)

- If you are using a three-person crew and a single adult victim is located, the entire crew must leave the building to remove the victim. One searcher can physically remove a child, and the searcher and victim can be led to a point of egress. Then the oriented officer and the other searcher can return to continue with the search. In that event, the second searcher should finish the search in the area where he or she was working when the victim was located, then go to the room where the victim was located to finish that room.

Multi–Family Occupancies 19

There are many terms used in the fire service today that have different meanings in different parts of the country. The intent of this book has never been to create conformity in terminology or to otherwise unify the terms used in our occupation. So as to provide an understanding of what I call things relating to occupancy type and construction, I will define the following terms related to multifamily occupancies. If my term or word doesn't meet your definition, then simply substitute your term for mine. What is most important is if you call any three-story apartment building in your community a garden apartment, then every department that could respond needs to understand your terminology.

- **Apartment.** An individual residence contained in a larger building.

- **Apartment building.** The structure that contains smaller individual apartments.

- **Duplex.** A building that is divided into two identical units that are inhabited by two separate tenants with separate entrances for each unit. Some duplexes are built side by side while others are top and bottom unit buildings.

- **Four-family.** A two-story building containing four individual identical living units with two units on the first floor and two units on the second floor. The building itself usually has one entrance and each individual apartment is entered off of a common hallway.

- **Older apartment building.** A multi-story building containing varying numbers of individual apartments. Older apartment buildings generally have at least one main entrance off of the street or parking area. Each individual apartment is entered from a common hallway.

- **Garden apartment.** A modern low-rise building housing many individual units. They usually have considerable lawn or garden space.

For the sake of this text, I will primarily discuss older apartment buildings and garden apartments in this chapter. I am of the opinion that duplexes and four family apartment buildings are generally fought similarly to single-family residential occupancies as discussed in chapter 18.

Problems With Searching Apartment Buildings

The main problem with searching apartment buildings is not having a good idea of the exact number of potential victims that could be located inside. Obviously some apartments are vacant, meaning that no human being is currently renting that specific apartment. Other apartment buildings may be unoccupied, meaning that the renter is not inside that individual unit at the time of the fire. Some apartments have restrictions on the number of occupants that are allowed in one unit, while others do not.

It is very difficult to pre-plan individual apartments. Hallways and shared areas such as laundry rooms and storage spaces can be inspected by firefighters. The same can't be said for occupied individual apartments. Although floor plans are generally consistent, furniture configuration and other items in apartments are not regulated in any form. Thus every apartment being searched can appear quite different from the last.

In newer garden apartment buildings, access to the building can be impeded by parked cars and other vehicles. Access to the rear may be difficult as well.

Automatic fire protection and alarm systems may not be present in these buildings. Occupants are not expected to always be in an alert state of mind. Sleeping individuals should be expected in these occupancies.

Older Apartment Buildings

Throughout this section, photos accompany discussions of each type of building. I consider older apartment buildings to be multi-story buildings containing a varying number of individual apartments. There is usually a front and a rear entrance to the individual building that leads to stairways which, in turn, lead to individual floors. At the top of each stairway is a landing that leads to a hall. This provides a pathway to each individual apartment.

Older apartment building construction 101

Older apartment buildings are constructed of either ordinary or wood frame construction.

Exterior load-bearing walls

Ordinary-constructed older apartment buildings used brick or a combination of brick and CMU for their exterior load-bearing walls. To determine if the exterior load-bearing wall is composed predominantly of brick, look for header courses (fig. 19–1). Exterior load-bearing walls that are made predominantly of materials other than brick will not have header courses. The exterior brick service will be only a brick veneer. Brick can be laid in other methods than with header courses in order to tie the brick assembly together for stability. A running bond would be an example of such a method of laying brick. In the absence of some form of header courses, ordinary constructed older apartment buildings can have CMU, steel, or a combination of brick, CMU, steel, or wood for exterior load-bearing walls.

Fig. 19–1. Header courses are the rows of what appears to be ½ sized bricks. In this instance, there is a header course every 6th course.

Wood frame older apartment buildings can have either balloon or platform frame exterior load-bearing walls.

Roof assemblies

Roof assemblies will almost always be supported by wooden rafters. Rafters will be horizontal in flat roofs and diagonal at about a forty-five degree angle in gable roofs. Most of these buildings that were designed originally as apartment buildings will have flat roofs. The cockloft area contains the last set of ceiling joists below and the rafters on top. The space between is filled with air and possibly insulation. Some buildings have a scuttle hole to the cockloft, and others don't.

The final exterior weatherproof surface will be made of tar paper, tar and gravel, or a newer membrane roof over either wooden plank or plywood. If plywood is found, suspect that alterations have taken place on the existing roof. This is the common roof assembly for either an ordinary or wood apartment building.

A single-family house that was divided up for apartments will usually have a traditional gable roof assembly. Access to the attic area may be from interior stairs, from a scuttle hole, or could be nonexistent. In the attic space you can expect to find anything and everything. Those with floored attics can be used for storage or even another living space. Those without flooring will probably be used for light storage, but mostly contain air and some insulation.

Floor assemblies

True ordinary-constructed buildings have upper floors members laid into either joist pockets or sitting on a corbel (a shelf made of brick). Floors are supported by 2×8-in. or larger wooden joists. These generally run parallel to the shortest exterior walls, normally the front and back walls. This indicates that the sidewalls and any interior wall that runs perpendicular to the joists is a load-bearing wall. On top of the joists are old, wooden floorboards. Plywood is used to replace those old floors that have been removed, for whatever reason.

The last set of joists that support the first floor are set into pockets or rest on a stone foundation. Regardless of the positioning of the joists, the construction will generally be the same.

Wood frame buildings also have wooden floor joist systems. In balloon construction, the joists on the foundation are nailed to the exterior sidewall. Joists will rest directly on the sill and be nailed to a stud. The sill is the bottom portion of the balloon frame exterior wall, as opposed to the sill in a platform frame which rests on top of the foundation. On upper floors, joists rest on a ribbon (also called a *ribband*) and again are nailed to the stud. On top of the joists are wood floorboards.

Prioritizing Search in Multi–Family Residential Occupancies

When you encounter a working fire in an occupied apartment building, search takes a high priority in the assignments made. Loss of life in these occupancies can be severe. However, the exact time assigned crew members search apartment buildings is dependent upon several factors in these extremely dangerous fires.

The order in which search is prioritized at these fires is dependent upon staffing, training, and experience level. You must take into account both on-scene and responding firefighters, as well as each crew's specific staffing level.

For even the most abundantly staffed fire departments in major metropolitan cities, there comes a point where the sheer number of potential rescues outweighs and outnumbers the firefighters on the scene. Assuming the best of situations with two firefighters per individual rescue, even the largest fire departments would be hard-pressed to complete a rescue involving more than 20 civilians without first controlling the fire and commencing adequate ventilation efforts.

Search is a priority in any advanced fire in an occupied apartment building. Under most circumstances however, I advocate the following order of assignments at a working apartment building fire:

- **Attack.** Get a line on the fire!

- **Ventilation.** Commence aggressive ventilation efforts.

- **Search.** Begin search efforts in areas as close as possible to the fire where savable victims may be. If staffing at this time is adequate, then multiple search teams should be assigned.

After search teams are assigned, the incident commander should quickly judge the effectiveness of the fire attack. Then the IC needs to consult with the attack officer as well as using gut instinct. If you believe that the fire is still advancing, then additional crews should be assigned to assist in fire attack. If crews are not available, then a decision must be made between continuing with search in areas remote from the fire versus removing search personnel from remote areas and reassigning them to assist in the fire attack.

Daytime Search in Older Apartment Buildings

Attempting search in older apartment buildings is one of the most challenging assignments that can be given to a search officer. I say this for a number of reasons. Construction concerns, inconsistent renovations that aren't up to code, and unfamiliar floor plans are among these factors. This coupled with the sheer number of potential victims and the concentration required to maintain continuity of search truly make this a challenge for even the most experienced officer.

Standard search

The officer and crew enter the building following the hoseline that has been stretched by attack as the first assignment at this fire (fig. 19–2). Assuming an initial hoseline is stretched up the stairway to division two and through the door to the apartment to the right, the search officer would evaluate conditions in that fire apartment. After consulting with the attack officer, if search believes savable people could still be in the rear of the apartment of origin, the search crew proceeds to that area and begins their search.

Fig. 19–2. An apartment building with smoke and fire showing on division two

I would expect the sleeping areas to be in the rear portion of this apartment. If using a four person crew, the officer has two choices as how to conduct this search. If it is a three-person crew, the second option is the only viable one. I realize my first option may differ from the true definition of standard search, but it is a viable option in a fire with the potential for many victims and large areas to cover.

With the first option, the officer splits the crew into two teams of two and designates a team leader for the newly formed crew. The officer assigns each crew a bedroom to search, giving the team leader of the other crew instructions to wait in the hallway if they finish first. Each crew then searches their assigned area until complete. In a smaller typical apartment bedroom, one searcher moves to the left and the other heads to the right. They meet somewhere on wall three at the rear of the bedroom. At that time they exit together and move on to the next area.

Some of these apartments may have the living room in the front of the apartment, followed by the kitchen/dining area, then the bedrooms in the rear of the apartment (fig. 19–3). Others may have the kitchen and dining area at the rear. Regardless of the configuration of the apartment, the officer should ensure that the search is done as close as possible to the fire where savable victims could be, then work back until they are finished with the apartment of origin. Once the apartment of origin is completed, the search crew moves across the hall to the other apartment on division two.

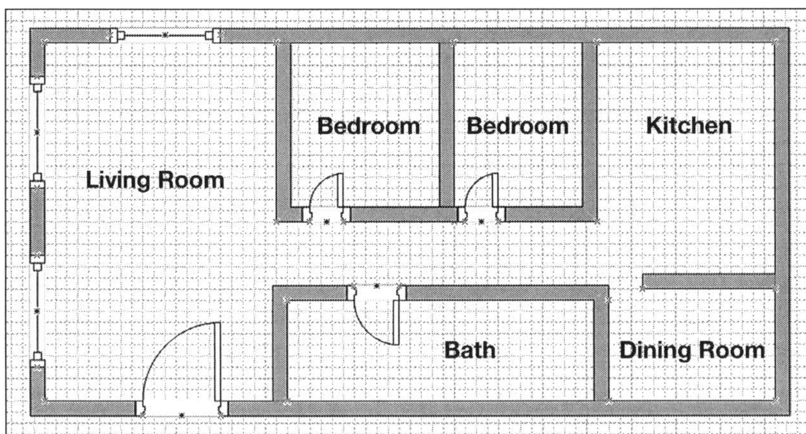

Fig. 19–3. Floor plan of a single unit

I would probably search the larger living room first. I would split the crew into two teams. A firefighter and the designated team leader head one way while the officer and other firefighter search in the opposite direction. Once the living area has been searched, move the team further into either the sleeping or cooking areas. Keep the search team divided, having one team search the kitchen while the other team searches the dining room, then proceed to the sleeping areas. Instruct each crew to search an individual sleeping area until the apartment is complete. At this time the officer should give the benchmark, "All clear on division two, going to division three," to the incident commander.

Depending upon the amount of air in their SCBAs, the crew moves up to division three to search the apartment above the fire apartment. When they've finished there, they need to move across the hall to the next closest apartment. Once those searches are complete, give the "all clear" and tell the IC that search is moving to division four. Again, this all depends on an adequate air supply. Assuming the low-air alarms are not sounding, we will repeat the search on division four and then return to division one for a quick sweep there.

In the second option, I would keep the crew together and search each room one at a time. There are two ways we could conduct this second option.

In the first way, if it is believed that savable people could still be apartment of origin in an area remote from the fire, then that is where the search officer should begin to search. Otherwise, the search starts in the apartment across from the apartment of origin on division two. A secondary search will be done in the apartment of origin at a later time. Remember, a secondary is a slow, methodical search to locate deceased victims.

The officer should maintain contact with his or her wall and have the other three searchers spread out toward the center of the room. They must maintain contact with one another, either physically or by using a tool or webbing. After the living room has been completed, the search crew moves back to cover the kitchen/dining area and then finish by searching the bedrooms and bath.

The above is the way a three-person crew, including the officer, would conduct the search, except two searchers would spread out towards the center of the room.

In the second way, the officer splits the crew into two groups. Two firefighters start a search in one direction while the officer and the other firefighter begin search the same room in the opposite direction. After the initial room was covered, they move through the entire apartment in a logical manner utilizing the same process detailed above the apartment search is complete. When the floor of origin search is completed, the officer announces the appropriate benchmark over the radio and moves the crew up to the division three. Search the apartment directly above the apartment of origin, then the apartment on the other side of a hallway. This process continues until they have finished search of the top floor of the building or they reached an area that was no longer IDLH.

A three-person crew, including the officer, can not be split safety. The paragraph above would not apply in that instance.

Team search

In this scenario I see no practical application of team search. The search rope used in team search is used to maintain an orientation with the exit. These apartment buildings have narrow hallways with easy access to at least one stairway. Individual apartment entrances are located along this hallway. Orientation with egress is usually not difficult in older apartment buildings. Although search ropes are an effective tool in maintaining orientation with points of egress in large wide open areas, they can become cumbersome and difficult to follow when navigating through individual apartments.

VES

The ability to use vent/enter/search in older apartment buildings is dependent on the height of the building and access to the bedrooms from the exterior windows. As in the building pictured above, access could be available to the basement level as well as divisions one, two, and possibly three.

A word of caution: ensure rooms that are being laddered for VES are actually sleeping areas. Attempting to search large living rooms, dining rooms, and kitchens using vent/enter/search can be an extremely dangerous task even if using the two-person oriented method. What makes VES effective and safe in sleeping areas is the size of the room and the fact that the room can be separated from the remainder of the building by closing the bedroom door. This may not be true of other rooms in an apartment or house. If it can be determined that a window actually accesses a bedroom, then VES can be employed as follows.

Two firefighters will ladder the bedroom window. The first firefighter up the ladder will clear the entire window and sound the floor. Upon entering the bedroom the searcher immediately locates the bedroom door to ensure that it is closed. The second firefighter should climb the ladder to the window and communicate with the searcher to verify their well being.

After closing the door, the searcher checks the bedroom, beginning and ending at the door separating the bedroom from the hallway. They may use either left- or right-handed search, depending on the environment and the searchers' preference. Once the search is complete, the searcher returns to the window and exits onto the ground ladder. At this point, both firefighters should descend the ladder and repeat this process at any other accessible bedrooms in the apartment building.

Oriented search

There are two methods of oriented search that could be used in this scenario. One is the oriented search and the other is the modified oriented search. Effective communication between the oriented officer and the searchers depends upon the size of the rooms. That is the sole factor in determining which method of search will be used. If communication becomes difficult, the modified method must be used.

When using the oriented search, the crew enters the building following the attack hoseline into the building. Upon reaching division two, the oriented officer will consult with the attack officer concerning conditions in the apartment of origin. Assuming conditions allow for search in the rear portion of the apartment of origin, the oriented officer leads the crew through the apartment to locate the hallway that leads to the sleeping areas of the apartment. The oriented officer will send one firefighter into the first room encountered. The searcher informs the oriented officer of the direction of search and, if possible, the number of walls in the room. At that time the oriented officer leads another searcher down the hallway to the next room and again sends in a single searcher to the room. That searcher also informs the oriented officer the direction of search and number of walls. If this is a four-person crew, the oriented officer will locate another room and send a third firefighter into that area to be searched. That last searcher also informs the oriented officer the direction of search the number of walls in the room, if possible.

At that time the oriented officer returns to the other searchers to check their progress. If time and conditions warrant, the officer should also monitor environmental safety concerns by checking temperatures at head and floor levels. They should also make a quick check back to the path of egress to ensure that it is still clear should rapid exit become necessary. When the first searcher has completed the assigned area, the oriented officer brings that searcher to the next area to be searched if one is available. In this case, they have completed the entire remaining area of the apartment of origin. At this time, the oriented officer will lead the crew out of the apartment of origin and across the hall to the next and last apartment to be searched on division two.

Understanding the layout of an apartment building such as this will determine how the oriented search is conducted. In this case, and as in the case in most apartments, the individual apartment will be treated similarly to the first floor of a two-story house. The oriented officer leads the crew in a direction that provides the shortest route between the door and the hall leading off of the living room. Once this point is reached, the oriented officer sends one searcher back to begin a search in the living area. Prior to starting the search, the searcher informs the oriented officer of the direction of search and the number of walls in the room. Then the oriented officer locates the next area to be searched. In figure 19–4, it is the hallway and two bedrooms between the living area and the kitchen.

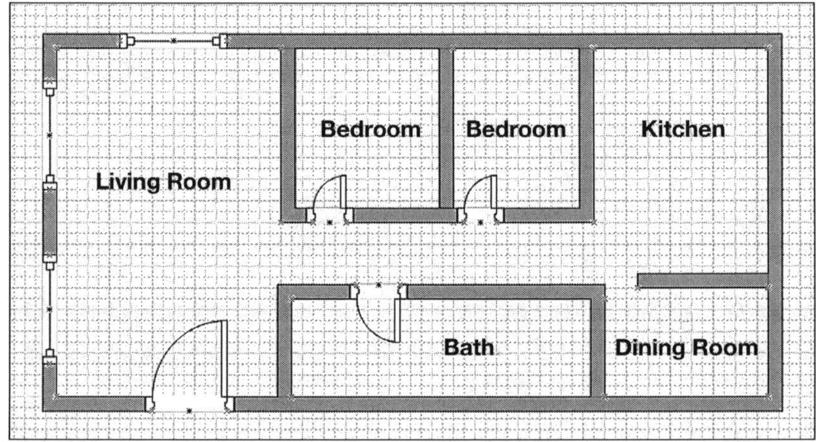

Fig. 19–4. Floor plan

The oriented officer then takes the other two searchers to each of the bedrooms and instructs them to begin search there. Prior to entry they must both inform the oriented officer of their direction of search and the number of walls, if possible. After that, the oriented officer returns to check on the progress of the first searcher. The officer may also check for safety conditions at that time,. then go back and check on the progress of the first searcher. Depending upon the condition of the original fire and attack's ability to control it, the oriented officer may opt to quickly check the conditions in the main hallway as well.

When the first searcher completes the assigned area, the oriented officer leads that firefighter to the next area to be searched. In this case, it will be either the kitchen or dining area. Upon availability of another searcher, they will be assigned to last area in the apartment needing search. At the completion of that final search, the oriented officer takes the crew back through the apartment by following his or her wall to the door. That should lead them back out of the apartment and into the main hallway on division two.

At this time the oriented officer gives the incident commander the benchmark, "All clear on division two, moving to division three." If the SCBA air supply is adequate, the oriented officer moves to division three and starts search in the apartment directly over the apartment of origin. When that is complete, the crew moves to the unit on the other side of the main hallway. Again, it is at this time the search officer will give the incident commander the appropriate radio communication, "All clear on division three, going to division four." If air content is still high enough, the crew moves to division four and completes that division as described above. Upon completion of division four, they move down and conduct a quick search on division one. The oriented officer gives the incident commander the benchmark, "All clear on division four, moving down to division one."

Modified oriented search

The modified method of oriented search is used when the area being searched is so large as that it will affect the ability of the oriented officer to communicate with the searchers. For this reason, instead of remaining in one location as the searchers search, the oriented officer is required to move into the area being searched as needed in order to maintain communication with the searchers.

It is best if the layout of apartments is known prior to the fire (fig. 19–5). The officer can then determine if the modified oriented search is required, prior to commencing the original search. Exactly how the modified oriented search is to be conducted depends on the size of the crew.

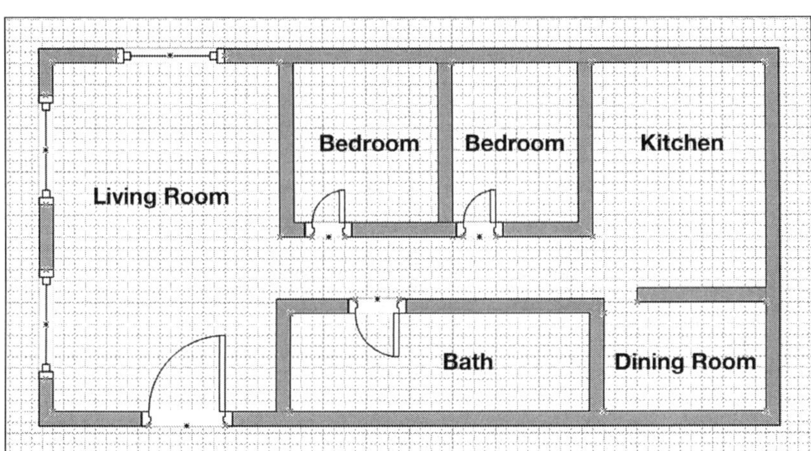

Fig. 19–5. Floor plan

If the crew has four members and the officer determines ahead of time that the modified oriented search is warranted, then the officer should split the crew into two search teams. Each team needs an oriented officer and one searcher. In the apartment building from figure 19–1, the oriented officer should have half the crew search an apartment on one side of the main hallway while and the officer and another searcher clear the opposite apartment on the same division.

The oriented officer moves into the apartment with the searcher and heads for what is believed to be the shortest route between the door of entry and the hallway leading to the other areas in the back of the apartment. Once that point is reached, the oriented officer will have the searcher begin searching the living area. The searcher must inform the oriented officer as to the direction of search and the number of walls in the room if it can be determined.

When that area is completed, the oriented officer leads the searcher to the next area. If, while searching that next area, communications become strained, the oriented officer should move nearer to the searcher's position in order to improve communications. When moving up, the oriented officer should go in the same direction as the searcher and simply follow the wall of orientation to a location where communication is improved. When the searcher returns to that location where the oriented officer moved up, the oriented officer can return to the original location. In this instance that would be the hallway between the living room, kitchen, and dining areas. At that time, the oriented officer directs the searcher to the next area to be searched. This process will continue until the entire apartment has been searched. At that time the oriented officer and searcher will follow the oriented wall out to the main hallway and meet with the other search team.

This is the time the appropriate benchmark, "Search to command, all clear on division two, moving to division three," should be given. The process as described above will be conducted on all subsequent floors.

Searching Older Apartment Buildings During Nighttime Fires

If conditions dictate that possible victims are likely to be located in sleeping areas due to the time of the fire, then the oriented officer should lead the crew to sleeping areas first in each individual apartment. Those sleeping areas should be searched as described above. Upon completion of search in the sleeping areas, the oriented officer can begin search with the crew in the living and kitchen areas of the apartment.

Maintaining Continuity of Search

Maintaining search continuity can become a daunting task when working fires with potential victims in apartment buildings. The apartment in figure 19–6 (a repeat of 19–2) that we have used for this scenario is divided in two sections by a firewall splitting the building vertically down the middle. This fire-rated separation is in the same area as the downspout in the center of the building, between the closely spaced bay windows. The fire-rated separation divides this apartment building into two sections. Each section has an individual stairwell that runs from the basement to the top floor. Each individual floor has two apartments. Including the two basement apartments, each section has 10 individual apartment buildings for a total of 20 apartments in the entire structure. If we assume that each apartment is occupied by at least one or two individuals, you can see that the potential for rescue and its effect on continuity of search will be very demanding.

Fig. 19–6. An apartment building with smoke and fire showing on division two

Continuity of search is assuring that a search is conducted in a logical, uninterrupted manner. It entails knowing where we have searched, where we are searching now, where we will search next, and when the entire area has been covered. Once a victim is located, we must assure that the remainder of that individual apartment is also searched in an effective and efficient manner while also keeping duplication of search efforts to the barest minimum. The ideal situation is to:

- Locate a victim.

- Have the searcher locating the victim remove the victim to the hallway (with assistance from the oriented officer if necessary).

- Pass the victim of to a rescue group.

- Have the searcher move back to the exact area where the victim was located to continue the search.

I realize this rarely happens. In order to maintain continuity of search, the oriented officer must get as much information as possible from the searcher regarding the location where the victim was found without delaying the victim's removal. Absent the original searcher returning to the apartment after passing off a victim, the oriented officer should request additional searchers from the incident commander. If that does not occur, when one of the other searchers from the crew finishes their area, the oriented officer should take them back to the room where the victim was located. The new searcher is then provided with as much information as possible in order to properly complete search in the room where the victim was located. From that point, the search can be continued has described above. We have done our best to maintain continuity of search.

Searching Newer Apartment Buildings

For the purposes of this chapter I will refer to the apartments in figure 19–7 as garden apartments. For the most part, garden apartments are associated with the following generalities:

- There are four individual apartments per floor.

- There are 12 apartments per section (four per floor by three floors). These 12 apartments make a section. Often, two or more sections are joined or abut one another. Ideally, sections are divided by a firewall or other maintained fire rated separation.

- There is usually more than one similar building in the area. These complexes can consist of as few as two or as many as a few dozen structures.

- Mid-sized to large complexes (four or more buildings) can be served by a single driveway and associated group of parking lots.

- Individual living units are confined to one floor, with the exception of town houses and some row houses.

- Individual balconies are typical.

Fig. 19–7. A newer two-story apartment building

Building Construction 101

Regardless of the exterior facade, the majority of garden apartments are wood construction similar to the construction of single family residential structures. There can be some local differences. The year of construction may indicate which type of wood construction was used, but as rule of thumb they will be wood construction.

Exterior load-bearing walls

Exterior load-bearing walls are commonly constructed of wood frame. Some garden apartments built in the East in the 1930s and 1940s have exterior load-bearing walls of ordinary construction. Wood joints were tied to exterior walls by a wooden ledger that was bolted to the brick or by corbel, which is a brick shelf constructed in the wall itself. In some "Section 8" housing projects, the exterior walls are of ordinary construction with concrete ceilings and floors.

The majority of wood frame garden apartments were constructed after balloon framing became extinct (between 1940 and 1950). This means that most garden apartment buildings prior to the 1980s are most likely platform construction. However, there is a possibility that garden apartments built prior to 1950 may be balloon frame construction. Newer garden apartments are now being constructed of truss frame construction, which poses additional hazards for firefighters. Still, the majority of garden apartments have exterior load-bearing walls constructed of wood using platform construction.

Ordinary constructed exterior load-bearing walls can have painted concrete masonry unit as a finished exterior. They can also use brick veneer walls. Wood frame garden apartments can have wood lap, vinyl, aluminum siding, or brick veneer as typical exterior finishes.

Roof assemblies

Roof assemblies can be flat roofs of concrete or wood joist, although flat roofs are the exception and not the rule. The flat roof assemblies may be beam-and-rafter with sheathing and asphalt roofing in older garden apartments. Parallel-chord truss or engineered plywood I-beams with tar and gravel over plywood are common in newer complexes that have flat roofs. Some low income housing projects can have concrete roof assemblies with a membrane covering of rubber, tar, or stone.

Pitched roofs make up the majority of roof assemblies utilized in garden apartments. Some of the older garden apartments employ a true gable roof system with wooden rafters as beams and a single ridge board that supports the slopes. The rafters are normally 2×6 in. The ridge board is normally a 2×6-in. board or larger.

Newer roof assemblies are of truss roof construction. These are traditionally 2×4-in. assemblies. However, some 2×6-in. assemblies can be found, but they are rare.

Floor assemblies

Floor assemblies are typically of wood joist construction. The two exceptions are as follows:

1. In concrete constructed garden apartments like low income housing projects, the floor assemblies are constructed of concrete.

2. Newer garden apartments can have engineered wooden I-beam or truss beam construction.

However, today the majority of garden apartments have wooden joist floor assemblies. These traditionally will have ⅜-in. or ½-in. plywood with a variety of floor coverings. A word of caution extends to very cheaply constructed units, which can have particle or chip board as the sub-flooring. These will deteriorate rapidly in fire conditions.

Searching Garden Apartments During Daytime Fires

The newer garden apartment buildings are becoming an ever-increasing fire problem in the United States. The sheer number of these buildings, along with their construction features, makes them a challenge for even the biggest and best fire departments. With the increasing population in the United States, I can only see the number of these buildings continuing to grow at a fast pace.

As you can see in figure 19–8, these buildings tend to be symmetrical. If you enter an apartment and turn to the right, you will head toward the kitchen. If you move down the hall into the adjoining apartment and turn to the left, you'll also head for the kitchen. This is done in order to conserve money for the builder and owner. Plumbing voids and pipe chases can be grouped together in this fashion, making it more economical.

Figure 19–9 illustrates the following scenario.

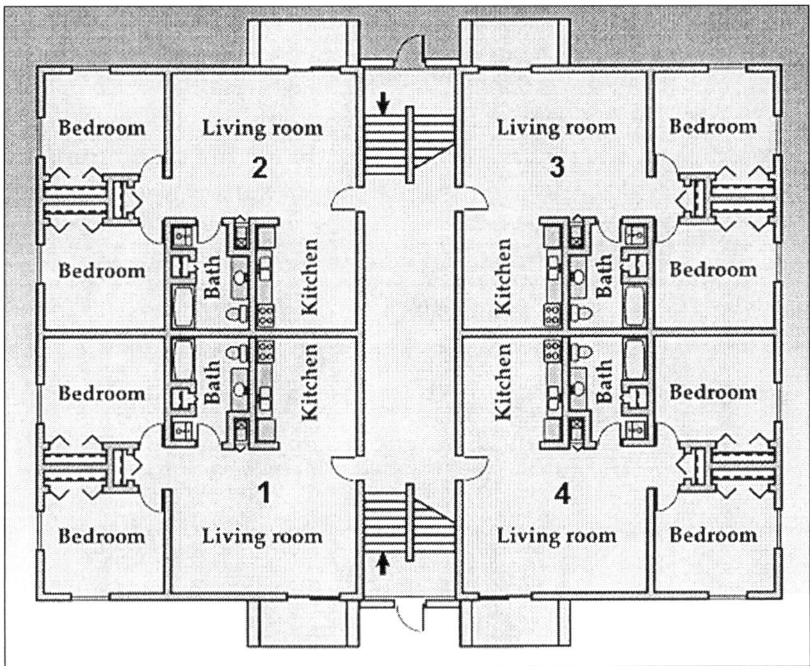

Fig. 19–8. Floor plan of a new low-rise apartment building

Fig. 19–9. A newer apartment building with fire and smoke showing on division two

Standard search

The search team should enter the building following the initial attack line inside. The search officer consults with the attack officer as to conditions in the apartment of origin. In this scenario, the attack officer has confirmed that the fire is in the living room area. The officer believes there could be savable victims in the rear portion of apartment number 1. The officer has two options for searching this apartment using standard search (see fig. 10–10).

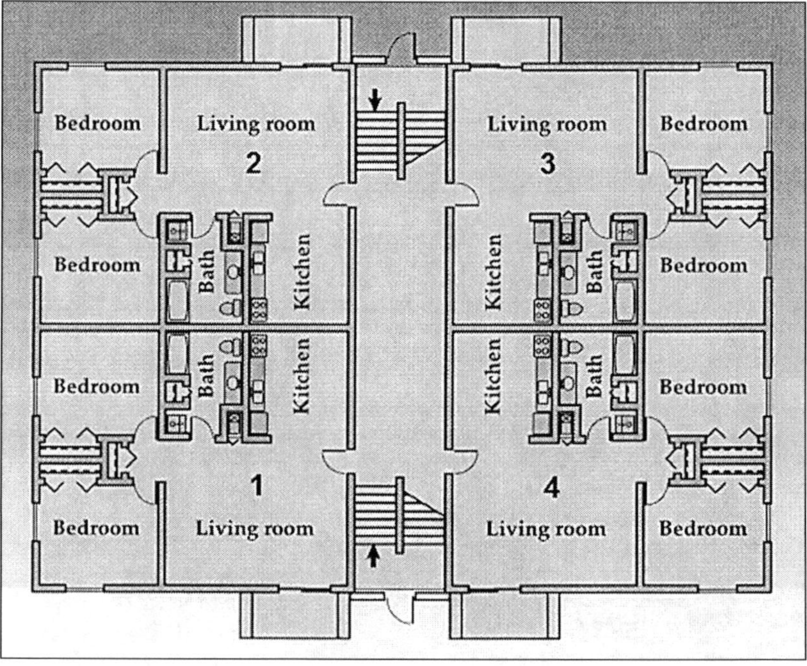

Fig. 19–10. Floor plan of a garden apartment building

In the first option, the officer begins a right-handed search, maintaining contact with the wall on his or her right shoulder. The other three searchers fan out toward the center of the room. The officer then proceeds forward along the right-hand oriented wall, and the crew searches the kitchen area. Upon completing the kitchen area search, the officer locates the bathroom and sends in one firefighter. They should quickly sweep the floor and tub, then exit the bathroom. Continuing to move to the right, the officer locates the next room. In this case it's a bedroom. Then the officer follows the wall and conducts another right-handed search in the rear bedroom. Depending upon fire conditions in the front bedroom, the officer may or may not conduct a search there at this time.

Upon completion of search, the officer directs the crew back to the main hallway on division two. The officer locates the door to apartment no. 2 and repeats the search in that area. If you note the layout of the first apartment, you should be able to perform the exact same search here but in the opposite direction. This time, use a left-handed search starting in the kitchen area. Then move to the bathroom, rear bedroom, front bedroom, and finally, living area. If you start a right-handed search, the order will be reversed.

The officer returns to the front entrance to start a search in apartment no. 4. Realizing that a living area is to the right, a right-handed search is conducted, starting in the living area and proceeding in the bedrooms. Finally, the crew will finish with the bathroom and kitchen.

Upon completion of search in apartment no. 4, the crew proceeds to a left-handed search in apartment no. 3. The reason the officer searches this apartment last is that it is the farthest distance from the fire. Once they've finished with no. 3, the officer gives the IC an "all clear" benchmark as the crew moves up to division three.

When the search crew enters division three, the officer should search in the same order as above starting in apartment no. 1, then going to apartment no. 2, then no. 4, and finishing in apartment no. 3. Once done, the crew exits apartment no. 3 and gives the incident commander the appropriate benchmark, "Search to command, we have an all clear on division three, moving to division one."

The second option that the officer has here is to conduct the search by splitting the crew with two members going to the left and two members going to the right. In this scenario, the officer would begin search in the apartment of origin by starting a right-handed search. Because of the configuration of the room, the crew should remain intact while searching the kitchen area. Once that is finished, one firefighter is sent into the bathroom to perform a quick sweep of the floor and tub. When they reach the bedroom, the officer splits the crew and instructs two firefighters to begin a left-handed search while the officer and other firefighter begin right-handed search. When they meet somewhere on the far side, they all exit the room together. A similar search should be conducted in the front bedroom if conditions indicate savable victims could be located there. If not, the case the officer leads the crew out the doorway into the main hallway on division two.

Next the officer leads the crew into apartment no. 2. The officer instructs two firefighters to begin a left-handed search in the kitchen area while the officer and other firefighter conduct a right-handed search in the living room. Due to the size of the kitchen and close proximity between the kitchen and bathroom, I would expect the search crew that began a left-handed search to continue on and search the bathroom. Upon completion of a living room search, each team of two would conduct a search in one of the bedrooms. Upon completion of search, the entire crew exits the apartment proceeds to apartment no. 4 where they would conduct a search in similar fashion. Upon completing search of apartment no. 4, the crew again heads to apartment no. 3. I realize that this option may differ from the true definition of standard search, but it is a viable option in a fire with a potential for many victims and large areas to cover.

Upon completing search of apartment no. 3, the officer gives the incident commander the appropriate benchmark and moves the crew to division three. The officer should conduct a thorough search of each apartment on division three in the same manner. Once this search is completed, the incident commander should be given the benchmark "All clear on division three, going down to division one." Then the officer takes the crew down to search division one following the same pattern.

Team search

I see no practical application for using team search in this scenario for the same reason I wouldn't use it in older apartment buildings.

VES

The same caution given in older apartments applies to newer garden apartment buildings as well. If it can determined that an exterior window services a bedroom area, then vent/enter/search can be utilized. In some instances balconies will lead to bedrooms.

In that event, two firefighters ladder the bedroom window or balcony. The first firefighter up the ladder will clear the entire window and sound the floor for sturdiness. If it is solid, the searcher enters the bedroom and immediately locates the door separating the bedroom from the hallway to make sure that it is closed. The second firefighter should climb the ladder to the window and communicate with the searcher to ensure their wellbeing. If the bedroom is accessed off a balcony, then the second firefighter or oriented officer should be positioned outside on the balcony at the door. After closing the door, the bedroom needs to be searched beginning and ending at the door to the hallway. Once the search is complete, the searcher should return to the window or balcony and exit. Then both firefighters descend the ladder and repeat this process at any other bedrooms in the building.

Oriented search

There are two variations to the oriented search that can be useful in this situation. The type of oriented search is dependant upon the size of the apartment and rooms being searched. The key indicator in determining which method of search can be safely and effectively used is the ability to communicate between searchers and the oriented officer. If the oriented officer can remain in the hallway and have an individual searcher conduct a search in each apartment while maintaining communications, then the first method is ideal because it allows the crew to search up to three apartments at one time. If communication cannot be safely maintained, then the second option or the modified oriented search is the only option.

In the first variation, the oriented officer leads the crew to division two where they commence search. If the attack officer advises that savable people could be in the rear of apartment one, the oriented officer instructs a searcher to begin in apartment one using a right-handed search that begins in the kitchen. I realize that sending a firefighter into an apartment involved in fire alone sounds dangerous, but you must remember that there is also an attack crew in that area with a hose stream between the searcher and the fire.

While performing this evolution, the oriented officer must inform the attack officer that there is a searcher in the apartment with them. The searcher will confirm the right-handed search. The officer will then proceed to the next apartment to be searched with the rest of the crew. Upon locating the next apartment, the search officer sends the next member of the crew in to begin search. The searcher will inform the oriented officer of the direction of search. Then the oriented officer proceeds on to apartment no. 4 to have the last searcher in the crew begin checking that apartment for viable victims.

At that point, the officer must return to apartment no. 1 to check on the progress of the first member of the crew. After verifying that member's status, the officer proceeds to apartments no. 2 and no. 4 to ensure that the rest of the crew is okay. In this instance, this is generally done by the officer remaining at the door and yelling to the searcher who will reply by yelling back to the officer. I find it best to slightly pull the lower part of my facepiece away from my face briefly while yelling to increase volume and clarity. Departments that use voice amplifiers need not do this. Assuming the firefighter searching apartment no. 1 completes it first, the oriented officer will take them to apartment no. 3 to cover the last unit on division two.

Whichever searcher finishes next should be sent into no. 3 to assist. If the first searcher in apartment no. 3 started a left-handed search, then that second searcher assisting should conduct a right-handed search until they meet, completing the apartment. At this time, the oriented officer gives the incident commander the appropriate "all clear" and moves to division three. On division three each apartment needs to be searched by one firefighter in the same manner as division two. After that, the crew heads to division one and repeats the search again.

If the size of the apartment and rooms prohibits effective communication between the searcher and the oriented officer, then the officer must use the modified oriented search.

Modified oriented search

In order to compensate for poor communication, the modified oriented search must be used. Ideally the crew is a four person crew. In that event, the officer and a searcher can conduct a modified oriented search in one apartment while another two-person crew conducts a search in the next apartment.

If the attack officer advises that there could be savable victims at the rear of the apartment of origin, the search officer should instruct the firefighter to begin a right-handed search in the kitchen. After finishing that and a quick sweep of the bathroom, the searcher moves up to the rear bedroom and the oriented officer moves to a position at the rear bedroom door. Upon completing search in the rear bedroom, the oriented officer and the searcher would leave the apartment and move across the hall to apartment no. 4 to begin their search there. The other bedroom in the apartment of origin is not searched because of intense heat and fire conditions. A secondary search in this bedroom will be conducted at a later time.

While this is going on, the other search team is conducting a similar left-handed search in apartment no. 2. The only difference between the searches is that this second crew would also have to check the second bedroom. Once the second crew finishes their search in apartment no. 2, they move on to no. 3 across the hall and repeat the process.

Once apartment no. 4 is finished, the officer and searcher move toward apartment no. 3 and check on the progress of the search being conducted by the other team. Depending upon their progress and location, they would either assist in the search or just await completion of the search by the second team. Upon completion of search in the last apartment on division two, the officer gives the IC the appropriate benchmark and the entire crew moves to division three. A similar search should be completed there, followed by another on division one.

Searching Newer Apartment Buildings During Nighttime Fires

If the time of fire dictates that any occupants are likely in the sleeping areas, then the searchers should begin there first. Whether an oriented search or modified oriented search is conducted, when searching at night time fires, consideration should always be given to the sleeping areas of individual apartments. Do not rule out the "oriented VES" if access can be made to bedrooms from outside windows.

Maintaining Continuity of Search

Maintaining continuity of search in apartment buildings can be very challenging and stressful for a search officer. Ideally rescue groups are established at every fire in a structure of this size. Once a victim is located, the searcher should pass off the victim to rescue and return to the area where the victim was located to continue the search. Absent that, the searcher that locates the victim must inform the search officer of the direction and wall they were using. When another searcher is available, that information needs to be passed on by the oriented officer so that duplication of efforts doesn't occur and no area is missed.

Low–Rise Motels 20

I have no statistical data to back this up, but I believe that the trend across United States is in building and operating low-rise hotels and motels as opposed to the 15- or 20-story grand hotels from the past. Large metropolitan areas like New York, Chicago, and Los Angeles have many high-rise occupancies in their city, but the rest of us are seeing a trend toward building and operating low-rise hotels and motels.

Problems with Searching Low–Rise Motels

Most of these occupancies will probably be protected by automatic sprinkler systems, which should reduce or minimize the likelihood of a working fire. This is especially true for anything built after 1975. The one exception is motels that are two stories and have the unit door exiting directly to the exterior.

The biggest problem associated with searching this type of occupancy is the potential for a high number of victims. Motel personnel should be able to give an indication of the number of rooms that are occupied as well as the likely number of guests. However, they have no way to monitor when those individuals are actually in their motel rooms, making it almost impossible to account for all registered guests. Then you have to consider that some hotel guests may have other people visiting them. Additionally, you have to be aware that guests in motels may not be awake and alert at the time a fire occurs.

Building Construction 101

Exterior load-bearing walls

New low-rise hotels and motels are wooden platform or truss construction. Some newer ones will be of ordinary CMU wall assemblies or lightweight concrete on metal deck with metal stud walls. One- two- and three-story occupancies generally have a center court or wing layout. High-rise hotels and motels use steel, concrete, or a combination of steel and concrete for their exterior load-bearing walls. To my knowledge, there has never been collapse of a modern high-rise hotel or motel in the United States due to fire. These buildings are built to withstand the effects of fire. However, as we all know, severe interior damage can result.

Roof assemblies

New low-rise hotels and motels have flat truss roof assemblies. They will use either tar and gravel roofs or membrane roofs. Gable roof assemblies in the same type of occupancy have truss assemblies with asphalt shingles or terra-cotta tiles. Some have resistive shake shingles. Modern high-rise buildings commonly use concrete or steel roof assemblies with built-up metal deck roofs or membrane roofs.

Floor Assemblies

New low-rise hotels and motels are built on a concrete slab. Those with basements or lower levels have wood floor assemblies. They most often use truss or engineered I-beam construction. Plywood sheathing can be present and carpet will finish off the floor assembly. Some low-rise hotels also have poured concrete over q-decking as floor assemblies. High-rise hotels and motels may use poured concrete over q-decking or just all-concrete floors.

Prioritizing Search at Hotel and Motel Fires

The hotel in figure 20–1 has 180 guest rooms. You do the math. Even at around 25% occupancy, we are talking 45 to 50 rooms that are occupied. Some rooms may house more than one guest. Search becomes a huge priority at these occupancy types. It actually becomes prohibitive due to the sheer number of firefighters required to locate and remove occupants to areas of safe haven. I mentioned safe haven because, unlike a single- or multi-family occupancies where the entire building becomes untenable, it may not be beneficial to remove located victims to the outside of the hotel. Assuming adequate space below the fire, it may be best to relocate displaced occupants one floor below the traditional fire department staging area. That would actually be two floors below the fire. It will take discipline for the firefighters that locate victims to take them down two or three floors below the fire. You will also need a complement of police officers to corral and control the occupants once they are brought down to the safe floor.

Fig. 20–1. A high-rise hotel with 180 rooms

Head counts are an effective tool in determining the scope of the problem at hotel and motel fires. Anything that can be done to keep displaced guests in one area and under some form of control allows for a more effective and efficient head count. This takes less time and effort than if the incident commander allows displaced guests to leave the property for whatever reason.

In the second sentence in the paragraph beginning this section I mentioned math. In my opinion, math plays a huge factor in determining what fire crews can actually accomplish at any major fire in a hotel or motel. Make two general assumptions: 1) that the built-in automatic sprinkler system failed and 2) that this fire has the potential for rapid spread due to highly combustible furnishings. Using these givens, a prudent chief officer can begin to grasp the number of firefighters required not only to extinguish, but to ventilate, search, and provide BLS and ALS to affected guests. Assume an absolute minimum of 50 potential victims if you have 50 occupied rooms. In reality, that number will probably climb to somewhere around 100 endangered guests. Then assume two firefighters per guest simply for search and victim removal. At this point, the math should become overwhelming.

Searching Low-Rise Hotels and Motels

I would expect that the first three divisions in figure 20–2 would be relatively clear of smoke and fire. I would also expect that this building has mandatory, built-in, automatic sprinklers. However, apparently something went wrong. I would also expect standpipe connections in several locations throughout the building. In order to expedite hitting these standpipes, pre-planning is a must.

Fig. 20–2. A high-rise hotel showing smoke and fire on division four

Standard search

The initial assignment of attack should have taken high-rise or hose packs up to the division below the fire. This may provide the search officer with an obvious route of entry. I would assume the search officer wants to take the shortest and most direct route into the building which, in this case, would be the main lobby front doors. Prior to entering the lobby, the search officer has to contact the attack officer via radio to ascertain the location of the standpipe they hit.

With fire showing on the B side of the motel, I would stretch the initial hoseline from the D side and advance toward the B side of the motel. This accomplishes the first responsibility of an attack crew, which is to get a hose stream between savable victims or property and the fire. If you do one, you do both!

The search officer heads to division two toward the standpipe connection and stairwell that the attack officer indicated. This diminishes the likelihood that the search crew gets struck accidentally by attack streams. It also speeds the location of the fire area. The search crew should advance down the hallway toward the B side of the motel. Once the search officer determines the border between savable and non-savable victims, they turn and commence search. Search should proceed back down the hallway of division three, heading toward the end of the hallway on the D end.

Because of the size of the individual rooms (fig. 20–3), I would instruct no more than two firefighters to search each room together. If the initial search crew is a team of three, then I would send one firefighter into the bathroom area and have the other two searchers start in the sleeping area of the room. If you have a four person crew, then I would split them and have each team of two search one room at a time.

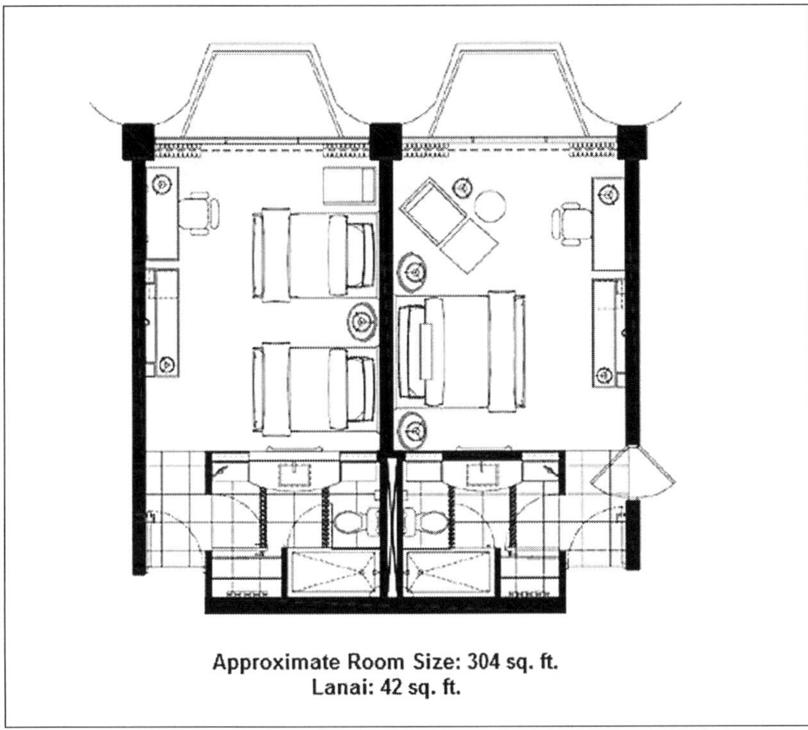

Fig. 20–3. Two individual hotel rooms with alternate layouts

After consultation with the attack officer, the search officer determines that rooms 332 and 333 to the B side of the motel has sustained such significant fire and smoke that no savable victims could possibly be in any of those rooms. Attack will do rapid cursory searches as they knocked down fire.

There are two variations of the standard search that could be conducted moving down this hallway, depending upon the size of the crew.

- **Three-person crew.** The search starts in room 330 with one firefighter entering the bath area while the other two firefighters move into the sleeping area. One goes to the left and one goes to the right. They will end up meeting somewhere in the middle, probably along the far wall of the room. After completing search in room 330, the crew should move across the hall to 331 and conduct a similar search. Upon completing search in room 331, they need to move out in the hallway and hit the next two rooms on either side. The crew should continue advancing toward the D side until the entire floor has been searched. The search officer will then give an "all clear" on division three to the incident commander and state that the crew is moving to division four.

- **Four-person crew.** Split the four person crew into two teams. Have one team search room 330 and the other team search room 331. One firefighter enters and begins a right-handed search while the other does a left-handed search. They meet somewhere near the back wall of the motel room. Upon meeting they exit the individual room together. I realize this violates the definition of the standard search, but in these extreme circumstances it becomes a viable option. The close proximity of the search officer makes it less of a risk. After completing a room, the crew moves down the hallway into the next room. They continue searching in this manner until they have completed division three. As in the previous example, the officer must give the appropriate benchmark and move to division four.

When moving up to floors above the fire, the search officer should enter division four from the stairwell on the D side. Then advance down the hallway counting doors until you determine the area to commence the search. Depending upon building construction and fire and heat conditions, I would probably start at the far end of the hall on division four. In this case, that's room 439.

Team search

I would not expect a search team to use team search in this occupancy type. The reason team search is used is to maintain orientation with your point of egress in large areas. Although this occupancy type certainly is a large building, the individual search areas are relatively small, between 300 and 325 sq. ft. Each of these individual search areas is fed from the main hallway. The point of egress to each hallway is a stairway at the end of the hall. While necessary in some occupancy types, the use of rope and team search tends to slow down a search crew.

VES

Vent/enter/search can be employed in hotels and motels if access can be gained by ground or aerial ladders. Tower ladders for buckets are very effective for this evolution. I cannot remember being in a hotel or motel that did not have self-closing doors to individual rooms. Except for very old hotels and motels, I would expect the doors to individual rooms to be closed. Regardless if access is made by ground or aerial ladders, one firefighter should enter the hotel room while the second firefighter assumes the oriented position outside the window.

Oriented search

This occupancy type is ideal for the oriented search. The points of orientation in this instance will be the hallway and the number of doorways that the oriented officer encounters. As long as the officer remains in the hallway and maintains an awareness of the oriented wall and the number of doors passed, it is virtually impossible to lose orientation with the egress point.

Using figure 20–4 as an example, the oriented officer enters division three and orients to the wall on their right. The officer then counts the doors as the crew proceeds from the

D side toward the B side of the floor. In this instance, there are 15 rooms on the oriented wall side in the hallway that need to be searched. This will take the crew into room 330.

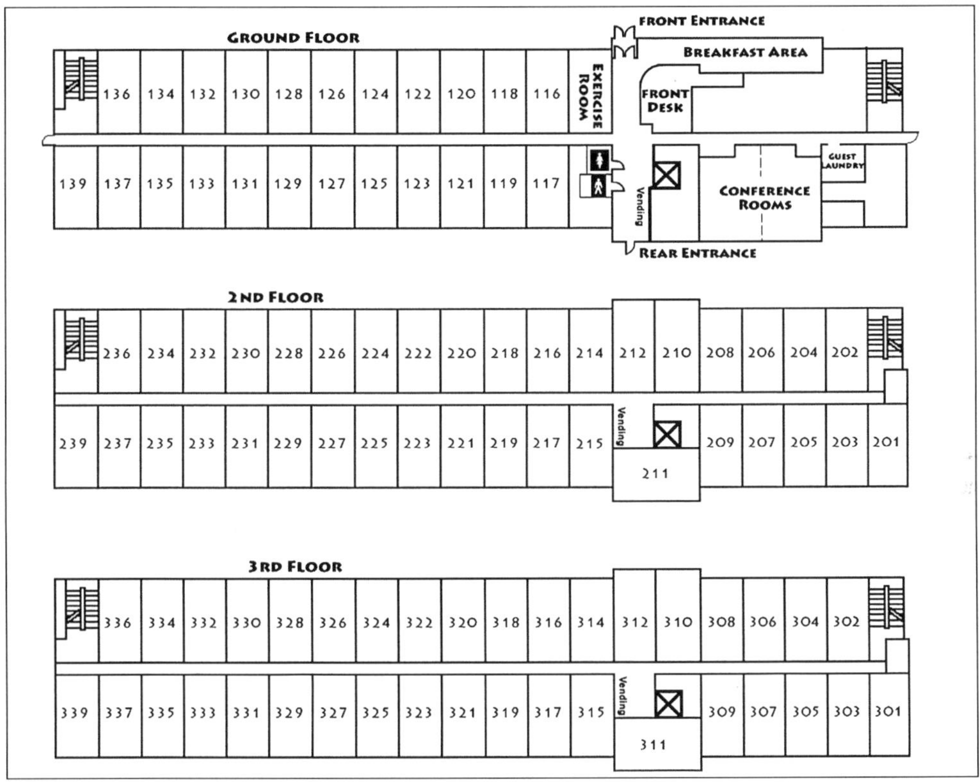

Fig. 20–4. Three floors of a hotel

As an example, if conditions become untenable or a victim is located after searching the fifth room on that oriented wall side of the hall, the oriented officer knows that the stairwell is ten doors back along the oriented wall.

- **Three person crew.** If the search team is a three-person crew, the oriented member leads the crew to the first room to be searched. In this case that is rooms 330 and 331. The oriented officer sends one firefighter into each of these rooms and instructs them conduct a search. Prior to entering, the searcher informs the officer of the direction of search. Once both searchers have commenced, the officer needs to move back and forth between the two rooms communicating with the searchers. The officer also needs to constantly monitor for changing fire conditions in the hallway. Upon completing the room searches, the firefighters exit their individual rooms and the oriented member leads them to the next room to be searched, in this case toward the D side of the building.

- **Four-person crew.** If using a four-person crew, the oriented officer will have three searchers available. A firefighter will be sent into each of the first two rooms to be searched. In this case, they enter rooms 330 and 331. The oriented officer then leads the last searcher down the hall to room 328 or 329 to begin search there. The officer then moves back toward 330 and 331 to check on the progress of those searchers. If time permits, they will move back to the third searcher in 328. Once an individual searcher has completed their room, the oriented officer moves that searcher down the hallway toward the next room to be searched. This process should continue until division three is completed.

Regardless if it is a three- or four-person crew, the oriented position will give the benchmark, "All clear on division three, moving to division four," when a particular division is completed. The oriented member should begin search on division four and all subsequent floors above the fire.

Maintaining Continuity of Search

Continuity of search will be extremely challenging at a working fire in an occupied motel. One of the best training sessions I ever conducted was in a vacant hotel in the Seattle, Washington, area. I began the training session by taking crews up to one of the divisions in the hotel. I started them searching near a stairwell. I had them search using the method that they currently employed in their department. I put a rescue mannequin further down the hallway in one of the rooms. When they located the mannequin, I instructed the crews to remove the victim just as they would today in their home departments.

When they brought the victim down the hallway and gave it to me in the stairwell, I instructed them to go back and pick up the search where they'd left off. They couldn't do it! Everyone was so focused on searching that they had completely ignored where the victim was located and what rooms had already been searched.

I then took them to the classroom and we discussed the oriented method of search. I talked about the position of the oriented officer. I told them how it was the oriented officer's responsibility to know what areas had already been searched and which had yet to be covered.

After that I took them back to the search area and did the same drill again. This time, they used the oriented method. Every time the crews located and removed a victim, they returned to the room where the victim was located and continued with the search. Not one room was missed, nor were any searched twice.

In real life, maintaining continuity of search in a low-rise hotel or motel will be a challenge. In my mind, the key is to establish a rescue group. When a victim is located, the searcher and the oriented member can remove the victim to the stairwell and pass it off to the rescue group. Then the searcher can return to the exact location where the victim was located and continue on with the search.

Nursing Homes and Other Health Care/Institutional Occupancies

21

It may seem odd to some, combining health care and other institutional occupancies with nursing homes. Jails and mental institutions are not physically set up the same as a nursing home. However, the basic principle behind organizing and conducting the search is primarily the same with nursing homes, hospitals, and other institutional properties.

Defend in Place

Defend in place is the best strategic option in these occupancies. Defend in place can be taken quite literally when defining the strategy. In essence, efforts are made to defend or protect victims suffering from the effects of fire and its by-products by leaving them where they are. There are far too many victims with far too many restrictions to make removing them all from the fire area practical. In nursing homes, many victims are bedridden with life-sustaining appliances connected to them practically around the clock. Individual living areas in nursing homes can contain from a dozen to more than 30 victims per wing.

Locks, doors, and bars that prohibit detainees from leaving prison facilities also slow down or completely nullify firefighters' ability to safely remove victims. Additionally, some victims could not be trusted one-on-one with a firefighter. We're not about to train firefighters in maintaining custody of their convict victims. Cell blocks and dormitory areas can contain dozens upon dozens of potential victims.

It is for these reasons that attempting to locate and remove individual victims in working fires in these occupancy types is extremely prohibitive. With two rescuers per victim in a single wing of a nursing home, it might require as many as 48 rescuers or more to locate and remove the occupants of that wing. Those numbers are well beyond the initial responding capabilities of the vast majority of fire departments in the United States.

Pre-planning these occupancies is a must. This pre-planning must include input from the management of the individual property as well.

Problems with Nursing Homes and Other Institutions

The number of responders makes these occupancies virtually impossible to search effectively. An incident commander must also consider the restrictions on the occupants of the facility. You just can't go into these places and yank victims out. In many cases, it will be physically impossible to do so or you could do irreversible harm to the victim.

Another concern is what to do with them once you remove them. Many of these individuals are frail and weak. Exposing them to extreme temperatures or inclement weather may bring on other life-threatening conditions such as pneumonia.

Building Construction 101

Nursing homes

There are four basic types of buildings that are used as nursing homes. These types of buildings include:

- **One story wing buildings.** These are modern one-story buildings built in the form of a hub with spokes. The center of the building contains cooking facilities, offices, and the nurses' station. Wings then jet out from this hub that contain individual living areas for the patrons (fig. 21–1).

Fig. 21–1. A wing-type nursing home

- **Converted buildings.** Prior to code restrictions in many Eastern cities, single-family homes in older areas were converted to nursing homes. The existing living areas generally remain in use as living areas for daytime use by residents. The kitchen area remained a cooking area used to feed residents. The sleeping areas were usually subdivided into many small, erratic individual sleeping areas for residents (fig. 21–2).

Fig. 21–2. Many buildings such as this have been converted into nursing homes.

- **Low-rise and high-rise facilities.** These newer, more modern facilities have living and eating areas on division one with individual rooms on upper divisions (fig. 21–3).

Fig. 21–3. A low-rise extended care home for the elderly

Exterior load-bearing walls

One story, wing-style nursing homes are normally made of wood frame or ordinary construction. As such, exterior load-bearing walls will be wood, block, or brick. These are often laid with wings that meet in a center hall area. Some are laid out as a square with a courtyard in the middle. Interior load-bearing walls are most often made of wood stud or concrete walls.

Converted structure nursing homes can have several types of exterior load-bearing walls. Many in my area have true balloon frame exterior load-bearing walls. Many are also old, very large, ordinary (brick) homes. Some of these are brick veneer over a wood frame, either balloon or platform. Interior walls are generally wood stud with lath and plaster or drywall finishes.

Low-rise nursing homes are made of ordinary construction or a combination of steel and ordinary exterior load-bearing walls. There are some three-story wood frame low-rise nursing homes. These are similar to garden apartments in structure, but not floor layout. They have larger rooms that are connected by a major interior hallway. The majority of these have all three floors above grade. Most are ADA compliant. Interior wall assemblies will be of wood stud, or in some cases, block.

High-rise nursing homes are found all over the country. These range in size from five to 25 floors and beyond. They can house hundreds of occupants. Some of these contain individual sleeping rooms off of a main hallway. Others have more of a hospital setting with individual wards that house several residents. These structures commonly have steel, concrete, or combination exterior load-bearing walls. Interior walls are generally drywall on steel stud or block (CMU). They tend to be very compartmentalized and produce hot fires.

Roof assemblies

Roof assemblies in wing-style nursing homes are usually constructed of wooden rafter or truss assemblies with plywood or chipboard sheathing and asphalt shingles. Trusses are predominant in newer buildings while older buildings have conventional rafter systems.

Converted nursing homes have predominantly wooden rafter roof assemblies with purlins or plywood sheathing. Exterior coverings are made of asphalt shingles with some older metal roofs or terra-cotta tile weatherproofing.

Newer low-rise nursing homes usually have wooden truss roofs with asphalt or tar and gravel exteriors. Some have poured-concrete roof assemblies. Some will also have steel bar-joist with metal deck or built-up roofs. Newer rubber membrane roof coverings are also out there.

High-rise nursing homes usually have roof assemblies of steel-bar joist with metal deck or built-up weatherproofing or concrete assemblies.

Floor assemblies

Wing-type nursing homes use either poured concrete or wooden joist floor assemblies. Some newer wing-type nursing homes may also have truss floor assemblies of either parallel chord wood trusses or engineered I-beam joists. Most of these wood floor assemblies have plywood sheathing and the majority have rolled or square linoleum floors for ease of cleaning. Concrete floors have a glaze finish or tile over the concrete.

Floor assemblies of converted structure nursing homes normally use wood joist construction. Most of these have tile floors over either plywood or true hardwood floors.

Low- and high-rise nursing homes have floor assemblies supported by steel bar joists. Some will be poured concrete supported by steel and columns. Bar-joist assemblies will support Q-decking with a thin layer of poured concrete. For sanitation and infection control purposes, carpet floor coverings are rare except in entrance areas. Most resident areas have tile or concrete floors.

Prioritizing search at nursing homes

In practical terms, search is not my most highly rated priority at a fire in this type of occupancy. The sheer number of potential victims and the restrictions placed upon removing them makes search virtually impossible. I believe that a defend-in-place strategy should be used where we:

- Initially confine and control the fire by the proper placement of a hose stream with sufficient volume to hold, if not totally darken down the main body of fire.

- Aggressively ventilate to remove the IDLH from the victims which, more likely than not, can be done much faster and safer with less staffing than commencing individual search and removal operations.

- Begin to protect and care for individual victims (after the first two steps are accomplished). This should be done by isolating the wing or area of origin from the rest of the building. Closing fire doors or otherwise isolating the area involved to keep smoke and other products from migrating throughout the building is the first step. Positive pressure ventilation via an exterior window or other opening while maintaining the original exit point is the next step. Sufficient crews should then be sent into the area of origin to begin locating and removing any viable victims. This evolution needs to be started as close to the fire as possible, then work back and away from the point of origin until viable victims have otherwise been protected or evacuated. Additional crews should be sent into adjacent wings and areas to tend to residents in adjoining areas. These victims should only be removed and taken to areas of safe haven if absolutely necessary. Air monitoring and additional positive-pressure ventilation keeping contaminated air from these areas is a must.

The above strategic and tactical evolutions for a nursing home can be considered for working fires in other institutional properties such as hospitals, long and short term health care facilities, mental institutions, and correctional facilities.

Searching Nursing Home Residential Areas

You may encounter small fires in these occupancies in which more traditional searches could be considered. In that event I offer the following.

Standard search

In the event a four person crew could split into two teams of two firefighters, each team can enter an individual residential area knowing that one bed will be to the left and the other to the right. Each firefighter should look for and remove one resident from each resident area, if physically possible. If not, the entire two-person crew will need to remove one victim at a time.

If you're using a three-person crew, two firefighters can enter an individual resident area as described above and the third firefighter can enter the next resident area, either further down or across the hall.

Team search

For the same reasons cited in previous chapters I do not believe team search would be advantageous in this occupancy type.

VES

If it is determined that we are actually going to conduct searches as opposed to using a defend-in-place strategy, then vent/enter/search may be considered as long as access to rooms via ground ladders or other means is available. Because of the number of victims and the limitations placed on these victims, vent/enter/search could become a very laborious task. Due to time constraints, I would consider utilizing VES only on division one of a nursing home and only in close proximity to the fire area. As in previous chapters, I wouldn't use VES in the fire room or room of origin. You should put the fire out first.

Oriented search

If and when you search, the oriented search is ideal for these occupancy types. One firefighter can enter an individual resident area to conduct a search. Due to the proximity of the individual resident areas and their orientation along a hallway or corridor, one oriented officer can easily handle two or three searchers.

The oriented officer should locate the area where the search is to commence. Then send a single firefighter into an individual resident area to begin search. If, because of the time of day, you expect each residential area will have two residents, you can send two firefighters into each individual resident area to facilitate search. That will allow completion of each resident area in a single search. If the crew is a four-person crew, the oriented officer should send two firefighters into one resident area and begin the next or adjacent resident area with the third searcher. Again, each individual victim may require two rescuers because of their size or other constraints. Adjustments may have to be made.

Once the room is completed, they should continue down the hall. The oriented member needs to keep track of the number of individual doors to resident areas that have been covered and their relationship to the means of egress.

Other areas

Most nursing homes and health care facilities have gathering areas where residents can socialize and conduct recreational activities with other residents. Additionally, these occupancies have kitchen and eating areas to provide meals for residents. These areas will need to be searched as well.

Searching Nursing Home Recreational and Food Service Areas

Standard search

Pre-planning is a must in nursing home facilities. It is best if, prior to commencing a search in recreational or eating areas in nursing homes, the officer in charge of the search has an indication as to the size and layout of the areas that will be searched. In figure 21–4, the recreational area is a very long and narrow room. I would envision chairs tables and at least one television located somewhere in the room.

The officer conducting a standard search in the recreational area might consider keeping the crew nearby and either starting a left- or right-handed search in the recreational area. The officer should maintain contact with the oriented wall as the searchers spread out laterally toward the center of the room. Webbing stretched between searchers and the officer spreads out the search some way away from the wall.

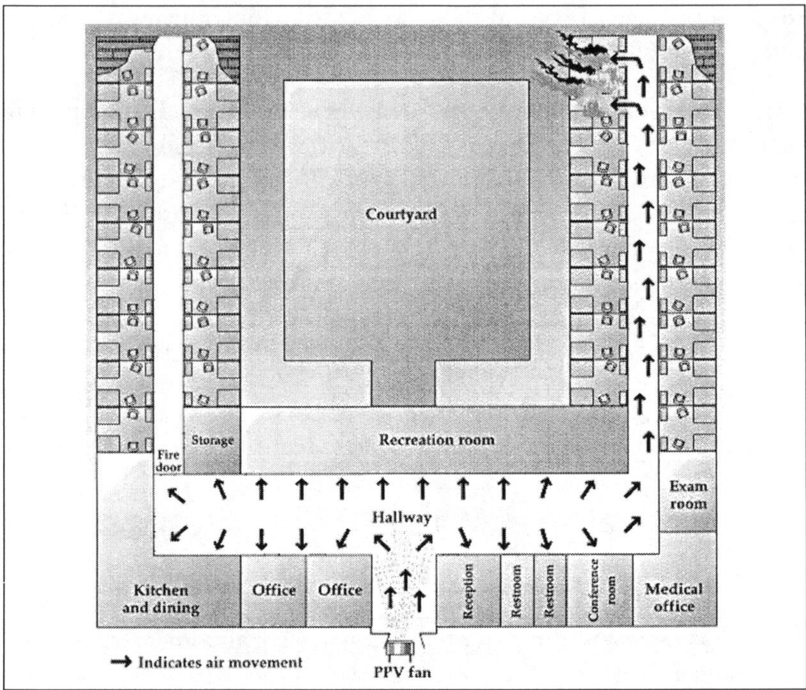

Fig. 21–4. Nursing home floor plan

If the officer of a four-person crew believes that voice contact can be maintained with another search team, the four-person crew can be split into two teams. Two members go to the left while the officer and another member go to the right inside the recreational area. Depending upon the size of the room, the officer can move out laterally three or four side crawls for the center of the room. The searcher can then move out three or four more side crawls from that position. This method will allow them to get closer to the center of the room, thereby covering more area.

The kitchen and eating area is L-shaped. I would envision a smaller kitchen area with a larger seating area. The seating area has many tables and chairs to facilitate dining for residents. As you'll find in chapter 22, searching areas that are predominantly furnished with tables and chairs is an extremely daunting task.

The officer opting to use the standard method of search probably would keep the crew intact. They might try to move around individual tables, maintaining coverage of the area as a whole. Again, this is an extremely difficult task.

My word of advice is to aggressively ventilate the area, removing IDLH conditions from the area and victims so that searchers can visually "look" for victims instead of just feeling blindly for them. In all likelihood this tactic will not only improve search conditions but will improve environmental conditions in the immediate area by removing smoke and carbon monoxide and replacing it with fresh air. If adequate ventilation and exhaust areas are provided in the above case, then the likelihood of pushing fire into unwanted areas becomes remote. This is especially true if a PPV fan is already holding fire to the affected wing, as in figure 21–4.

Team search

If the recreational area or kitchen and eating areas in this facility were empty of furniture, then a team search would be a very practical and safe method to use. With chairs, sofas, tables, and other furniture in the recreational area or dining tables and chairs in the eating area, I would bet any search rope will become tangled and rendered useless in this occupancy type.

Oriented search

Due to the size of the rooms encountered in kitchens and dining areas, the oriented search is not a practical or safe method of search unless a means is provided to maintain orientation with your point of egress. A 2½-in. handline is used in this scenario.

A dry 2½-in. handline is stretched from the exterior, either from the standpipe connection or an engine, into the area to be searched. In this instance, that is the recreational area or the kitchen and dining area. The line is stretched to a point adjacent to the point of entry and then charged. This line is not used to extinguish fire, but rather it is utilized to create a 2½-in.-high wall that the oriented officer can use to maintain orientation with the path of egress. Once stretched to an adjacent wall or as far as the line will go, the oriented officer should have the line charged. Then the officer needs to turn facing the path of egress and have the searchers work out laterally away from the line and then back, as illustrated in figure 21–5. Once both searchers return to the oriented member, they all move up 3 or 4 ft. and move repeat the process. This evolution will continue until they are back at the door to the hallway.

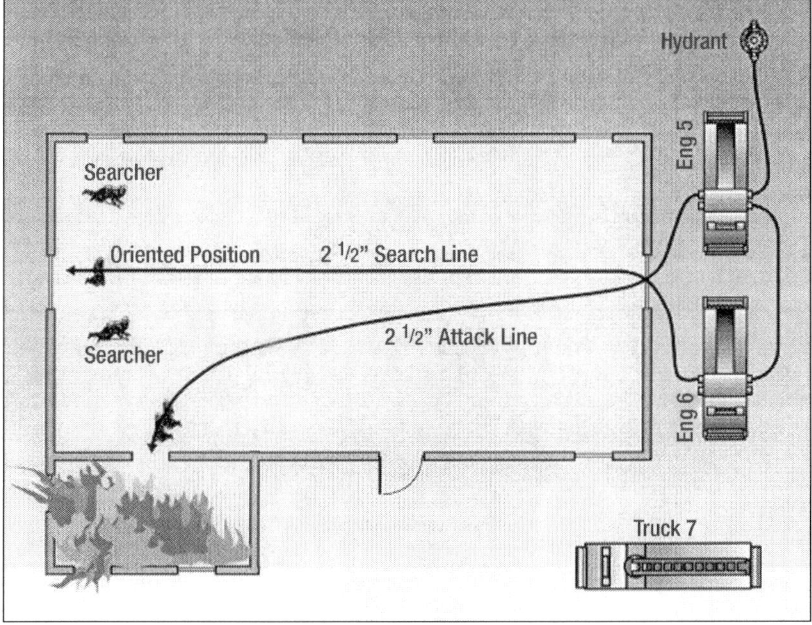

Fig. 21–5. Oriented search using a 2½-inch handline

Maintaining continuity of search

If you expect that residents will be removed from their individual living areas, then a rescue group will be a necessity. I caution against removing individual residents in these occupancy types, especially nursing homes and hospitals, because of life sustaining appliances like IVs and heart monitors that could be attached to individual patients.

The oriented officer must ensure when a victim is located that either the same searcher returns to continue where search was left off or another searcher is brought to that area and informed of the direction and progression of search. This is a process similar to the one discussed in low-rise hotel and motel fires. It is only through this disciplined process that we can assure continuity of search is maintained in these occupancies.

Restaurants

22

When I go to different fire departments across the country to teach, I often discuss search. When I discuss search, I usually throw in my two cents worth on searching restaurants. My basic line is usually something like, "Okay, in your department tonight, how many of you would search a *[insert restaurant name]* in zero visibility?"

I usually get a lot of blank stares.

Almost every fire district in the United States contains a restaurant in some form or fashion. Even in those very rare districts that don't have a restaurant in its jurisdiction, I would be confident in saying that a mutual aid response call would provide an opportunity for local firefighters to deal with a working fire in a restaurant.

As with the previous chapters discussing nursing homes and other institutional properties, you will find that there are a wide variety of building types that are used to house restaurants. A big concern is that there are so many possible combinations of construction types, from those originally designed and built as a single family occupancy to balloon structures to 20×20-ft. concrete block buildings. There are several old 10×20-ft. hamburger joints still in use in Toledo, although they aren't being used to sell food. They are CMU buildings with flat roofs and no basements. On the other end of the spectrum are rotating restaurants on the top floors of high-rise buildings. Elevator shafts, pipe chases, HVAC systems, and hundreds of patrons can complicate fire control and rescue efforts in these occupancies.

Problems with Searching Restaurants

Many problems present themselves when it comes to searching restaurants:

- Numerous construction types and methods
- Uncertainty regarding the actual number of patrons
- Lack of standardization in floor plans and layouts
- Movable tables and chairs
- Uncertainty about physical alertness in establishments that serve alcohol
- High fire loads creating large fires that hinder search

Building Construction 101

The construction type for restaurants can be as varied as the menus and patrons. They range from new, lightweight construction to old, converted mills. You will encounter everything from CMU 20×20-ft. fast food drive-throughs to multi-story, lightweight dining facilities, and all the varieties in between. Some will be protected by sprinkler systems while others may not even have hand-held extinguishers.

Exterior load-bearing walls

Exterior load-bearing walls can be made of almost any material. Older restaurants are erected using ordinary construction, wood frame, concrete, and even a few are made of steel. Converted buildings are used as restaurants, and renovations can be plentiful. Exterior load-bearing walls in newer restaurants can be made of steel, concrete, CMU, wood, or any combination thereof. A familiarization walkthrough inspection provides most of the construction information you'll need to operate at the next fire. This information needs to be updated at least annually. On top of that, check the type whenever you notice construction going on while you drive to and from runs or to the store. Know your buildings!

Roof assemblies

Roof assemblies can be of several types. Older restaurants have wooden plank roofs over wooden rafters. They can have tar and gravel or a rubber membrane exterior covering. Plywood can also be applied on top of the rafters instead of planks. Some older roofs can also be poured concrete.

Newer restaurants will probably have lightweight truss roofs that will support a variety of exterior coverings. Metal deck or built-up roof, asphalt shingles membrane roofs, or simple metal roofs will be the norm. Again, know your buildings.

Floor assemblies

Floor assemblies will normally be wood joist, wood truss, or concrete.

Prioritizing Search at Restaurant Fires

If you want to get into a discussion around the kitchen table at the firehouse, pose this question:

> *Let's say we pulled up at a working fire in a restaurant, maybe the size of an Olive Garden. We have heavy smoke showing with fire blowing from the rear of the restaurant. People outside are telling us that they think there are still victims inside. It's approximately 9:30 p.m., and the parking lot has 30 or so cars in it. Being the first engine company on the scene, we start to advance the hoseline through the main entrance and encounter zero visibility. After moving about 5 feet into the building, we locate a victim. Should we remove the victim or continue on to darken down the fire?*

This may seem like a silly question, but stop and think about what actually will occur if you abandon the fire attack and remove this first located victim. You may save one life, but what are the chances for any yet-to-be-located victims still in the structure? There are times when a fire crew must abandon tradition and develop a new strategy or tactic that is designed to do the most good. In this case, darkening down or extinguishing the fire may do more good for the most victims. It's hard to say, but I think the question, discussion, and possible answers are well worth the time taken to pose them.

The priority for searching a restaurant fire depends on several factors. If the fire occurs when no cars are in the parking lot at a restaurant, then the priority for search can be put down on your list of things to do. At the same time, the priority of search at The Station fire in Warwick, Rhode Island, would have been way down on my list of things to do. It cost 100 lives, but searching with that amount of fire present upon arrival could cost even more. Having a solid understanding of what your fire crews can actually accomplish, along with educated estimates of potential life safety problems, can bring you to an intelligent decision on how to weigh the advantages of search against other evolutions that need to be performed at a working fire.

The phrase "do the math" comes to mind. The number of firefighters required to locate, remove, and then tend to victims compounds very quickly. Considering that the victim/rescuer ratio is one victim to two rescuers at a minimum, you can run out of firefighters very quickly.

Searching Restaurants

The physical process of searching for victims in a restaurant is a daunting task, especially in zero-visibility conditions. You have to remember that if a search crew enters a restaurant (or any other building for that matter) with only 5 ft. of visibility, the amount of visibility should remain constant or improve as the search continues. Depending upon fire and ventilation conditions, visibility often decreases within the structure.

Take a look at figure 22–1. How would a single crew of three or four firefighters search around these tables and chairs while keeping a mental log of what has and has not been searched? Believe me, I have thought long and hard about how a crew could physically search a restaurant similar to this one and maintain an awareness of the progress made in the search. With so many identical items in the room, it's difficult to keep track of where you've searched, where you will search next, and most importantly, when you've completed searching the seating area. That is another good question to pose to your crew at the kitchen table.

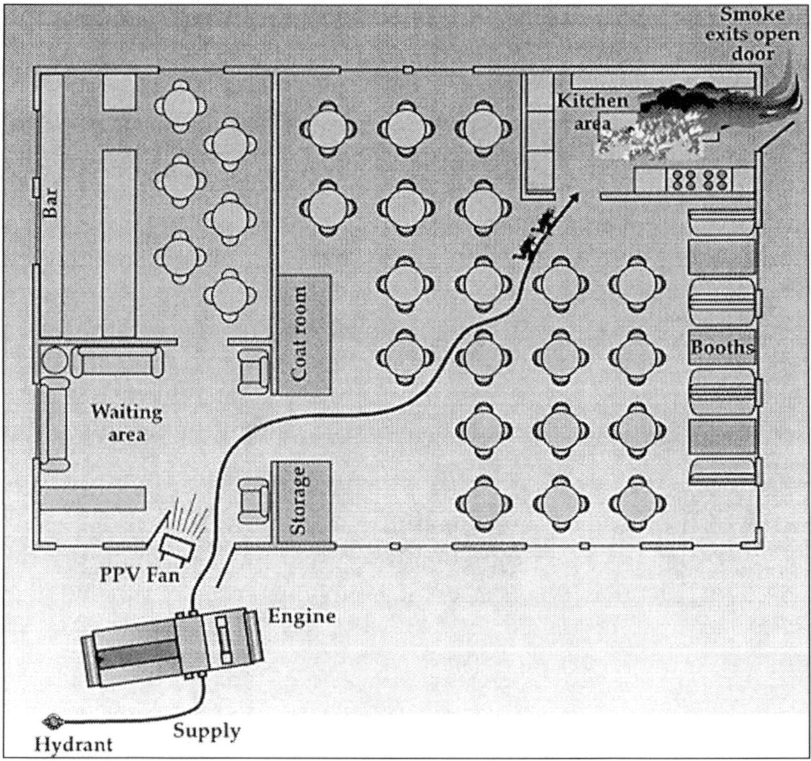

Fig. 22–1. A single engine working a kitchen fire in a restaurant

An even better suggestion would be to take your crew to a brand new restaurant in your district prior to their opening and ask the manager if you can simply look around for a moment. Explain that your intention is to discuss with your crew how you could conduct a search in their restaurant in zero visibility if a fire were to occur.

Another good kitchen table question is, "How long is too long?" Almost any crew in the United States could search a restaurant similar to the one pictured above, but could they do it within that 15-minute, single-SCBA bottle time frame? I don't believe a single crew, even those rare five and six person crews, could come close to searching a restaurant of that size and expect to still locate and remove viable victims in the last few moments of their search.

There is one last item for consideration prior to discussing individual search strategies. Take another look at figure 22–1. The tables and chairs in this illustration are not fixed but rather just sit on the floor. Most firefighters tend to emulate a bull in a china shop when they search. Our tendency is to move (or shove) things out of our way. In figure 22–1, after encountering the first two or three tables and knocking them around, the entire room will be a jumble. If you're envisioning a huge mess with piles of overturned tables and chairs, you're probably more correct than not.

Standard search

For many of the reasons explained above, any attempted search in a restaurant must be considered very carefully. In figure 22–2 there are 55 tables throughout the seating area. The entrance to the eating area is on the A side of the illustration where areas 5 and 6 are located. If we assume zero visibility throughout the entire area, you can quickly imagine the daunting task in maintaining orientation with your point of egress.

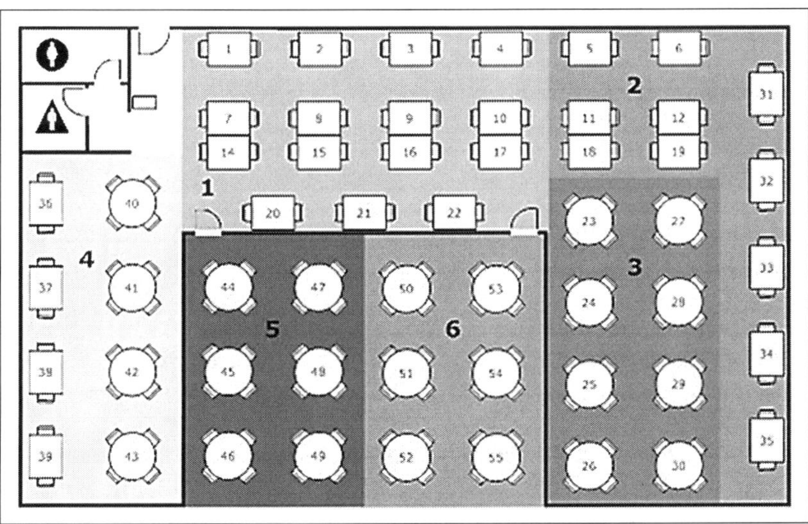

Fig. 22–2. A restaurant seating area with 55 tabletops

Using the standard method of search, the officer and crew would enter on the A side of the eating area. The crew must maintain integrity, meaning that they must remain within voice contact of one another. The only way to ensure orientation is to locate or create a wall. I will explain the latter in the oriented method of search section below. In this case, we will use the wall on the left side of section 5 running along tables 44, 45, and 46. Once the officer is oriented with the wall, the searchers should move out laterally, sweeping around, in between, and under tables. They continue until they reach either the wall on the right side of section 6 or begin to lose communication due to the distance between the officer and the searchers. The officer, once having located and oriented to the wall, should also move out laterally searching around, underneath, and between tables as well.

Once this section is completed and the officer and crew have returned to the wall, they should move through the door and either begin searching in section 1 or move to the left and beginning a search in section 4, depending on conditions. All members search a row of tables until they meet at the wall on the opposite end of the section. At that time they return to their wall. This process continues until all areas of the restaurant eating area have been covered. Hopefully the officer has located the other areas of the restaurant during the search such as the kitchen and prep areas as well as any bar and office areas. Those should be searched as they are encountered unless other information is provided, such as all workers are reported out of the building.

Team search

Team search is not an option in this occupancy type. Any search rope would quickly become tangled and ineffective in providing an avenue of egress from the area being searched. If the rope were pulled taut in frantic effort to exit the building, it would eventually lead to the area of egress. But in doing so, it could result in the entanglement of many tables and chairs. Those could become overturned and dragged into positions that might confuse and even prohibit ease of egress.

I know of no procedure using team search in which a search rope could be affixed high up off the ground so as not to become entangled with tables and chairs. Absent extensive preplanning and regular re-familiarization, I cannot envision anchor points to attach the search line to that would be 36 to 48 in. off the floor.

Oriented search

As my grandmother would say, "If I had my 'druthers,'" and I were to contemplate searching this type of occupancy, I would use the oriented method with the assistance of a 2½-in. hoseline that would create a 2½-in. high wall for the officer to remain oriented.

The oriented officer and the crew would stretch the dry 2½-in. line following the initial attack line into the building. Returning to my rule about where search should begin, this tactic will assure that the search crew gets as close to the fire area as possible using a safe avenue into the building.

Once the oriented member has determined where to commence search, they order the 2½-in. hoseline charged, making sure the bail is closed. Then the officer should literally sit on that line facing the direction of egress and instruct the searchers move out laterally. Searchers should be directed not to enter any area or room other than the area being searched without informing and receiving a confirmation from the oriented officer.

Once the searchers reach a wall or other obstruction, or communication is no longer affective, they need to move back to the oriented officer. At that time the oriented officer and searchers all move forward 3 or 4 ft. and each searcher would again move out laterally from the hoseline.

If the oriented officer and searchers return to the point of entry and realize there are additional areas that were missed, they should shut down the line. Then they need to bleed as much water out as possible by opening the nozzle. After that, they should drag the line into the next area to be searched. The line should be stretched to an adjacent wall or until all the line has been played out. At that point, the line should be re-charged and they commence another search as described above.

Maintaining continuity of search

Maintaining continuity of search is an extremely difficult process in restaurants. Whether using the standard method of search or the oriented method, there are few landmarks that can be used for orientation. Even the landmarks that are present, such as tables, are difficult to tell apart.

Unlike searching apartment buildings, hotels, and motels, there are no doors or individual beds that can be used to return to a specific area where a victim was located. The oriented officer can count tables, but in zero visibility and with all tables appearing identical, this would be a very arduous task.

My word of advice would be to have the initial searcher that locates a victim remain as close as possible to the area where the victim was located. Then have the rescue group move up to that location and begin removal of the victim. That way the searcher can then continue on with the search.

I realize that there are departments that say their firefighters would never pass off a located victim to another firefighter. That's really a shame. I certainly hope my loved ones never visit those communities where firefighters maintain that protocol. I do not believe any of us could state truthfully that we would want someone to search for, locate, and remove multiple members of their family that way.

A rescue group is a must at a working fire in a restaurant. The sheer number of potential victims and the amount of firefighters required to establish a rescue group could quickly grow into double digits. Ten firefighters provide you with five rescue groups. In these extreme situations where multiple victims are possible, five rescue groups could be kept extremely busy simply removing victims to the outside.

Once outside, EMS or patient care groups must be established to receive and begin ALS or BLS patient care on removed victims. BLS victims can be tended to with a rescuer to victim ratio of 2 to 1 in extreme circumstances. Under these extraordinary circumstances, ALS victims should be tended to with a minimum rescuer-to-victim ratio of 3 to 1. As you can see, the number of personnel required at these occupancy types can become very large. They will be needed in an extremely short period of time if they are to be truly effective.

Final Thoughts

Figure 22–3 depicts a newer, modern restaurant. It is made up of approximately 8,000 sq. ft. of dining and cooking area. Assuming zero visibility and an optimum crew of three searchers and one oriented officer, each searcher would have to cover approximately 175 sq. ft. every minute. In doing so, they would almost surely have to continue searching as their low air alarms are sounding to cover the entire area in a 15-minute time frame. That is the equivalent of a single firefighter covering a 14×13-ft. bedroom in one minute. Envision trying to cover that space while moving around and searching under tables and chairs.

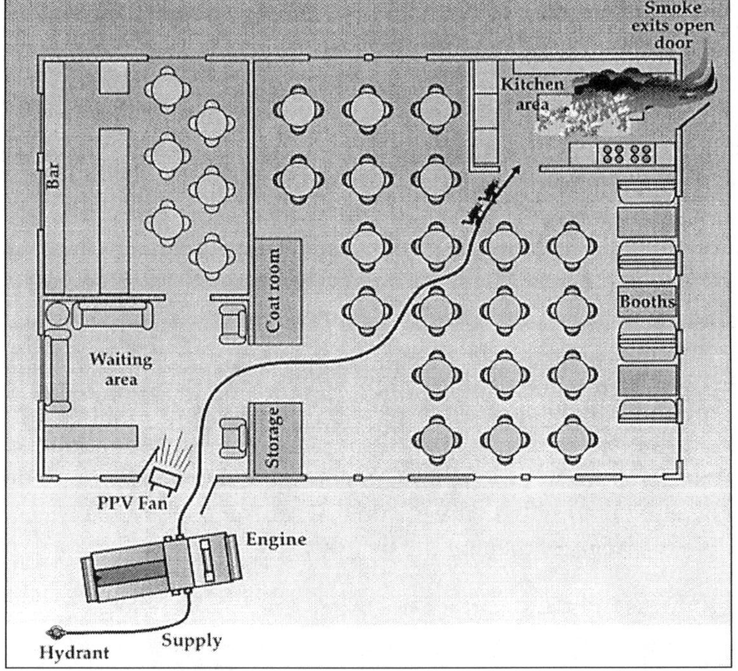

Fig. 22–3. A kitchen fire in a restaurant with both attack and search hoselines deployed

Most departments in the United States don't have the initial on-scene staffing to establish a water supply, begin a fire attack, and provide a crew for initial search in an occupancy of this type. In an advanced fire with zero visibility, I would not feel comfortable as an incident commander sending the initial attack crew inside a restaurant with anything less than a 2½-in. handline. The potential BTUs that could be produced would quickly overwhelm any line smaller than that.

To stretch a charged 200-ft., preconnected 2½-in. hoseline inside a restaurant will take a minimum of four firefighters and an officer directing the evolution. Considering the staffing required for this fire attack, the establishment of a water supply, pump operator, incident commander, and a three or four person search crew exceeds 13 firefighters. NFPA suggests an initial response of 13 as a minimum. If that is all that you have responding and you expect to be successful in darkening down the fire and locating viable victims, all 13 of these personnel will be stretched to their limits.

The only way I see around this would be similar to an aggressive defend-in-place strategy, coupled with positive pressure ventilation. That will remove the IDLH from the restaurant so firefighters aren't blindly feeling for potential victims.

I offer the following priorities at restaurant fires:

- Get a hoseline of adequate size between the fire and savable victims, then darken the fire as quickly as possible.

- Position a positive pressure fan or two in an appropriate location with an adequate vent hole behind the fire. This will increase visibility rapidly and dramatically while forcing the by-products of combustion out the vent hole and pushing fresh air into the restaurant area. That will lower temperatures at floor level to tolerable levels. In an 8,000-sq.-ft. building, I would expect visibility to go from 0 to 10 ft. within a matter of minutes.

- Send as many remaining firefighters in as possible to "look" for and remove located victims.

Strip Malls 23

Strip malls are a phenomenon that has occurred over the last 40 years or so. In essence, they are the same as taxpayers that sprung up throughout the United States in the late 1800s and early 1900s.

I have both good news and bad news regarding search in strip malls. The good news is that the threat to life safety in these occupancies is almost always very low. I will explain this a little later. The bad news is that these buildings are not constructed well and, as you will see, are extremely prone to early collapse. In fact, if we look at strip malls as they relate to firefighter fatalities, they are one of the most prolific and dangerous killers of firefighters today.

Always remember that these are disposable buildings. Once life safety is accounted for, I have no problem with an extremely conservative strategy centered on an exterior knockdown. No unoccupied bicycle shop, barber shop, or burger joint is worth the life or limb of any firefighter.

One last word concerning strip malls. There are two types of owners of strip malls: those who have insurance and those that do not. It makes no difference to me whether a store or building owner does or does not have insurance. Building owners make decisions concerning their wants and needs for insuring a property just as I make decisions as an incident commander to keep my firefighters safe when the only thing they are protecting is property.

Problems with Searching Strip Malls

There are two main concerns when it comes to searching a strip mall. First is the time factor. As you will learn below, these buildings are very prone to early collapse in a working fire, especially if the fire and heat are significant at the roofline height. The second main concern is simply the sheer number of combustible items in these occupancies. Fire loads can be significant, which may lead to significant and rapid heat and fire buildup.

Building Construction 101

Strip malls are the modern taxpayers. Taxpayers and their current brethren are both disposable buildings. Don't let developers fool you. These building aren't designed to take much more than strong gusts and the average snow load. "Fire" was probably not a word in the architect's dictionary. There are probably no engineered extras that will help if fire attacks the building. They are there to make money until something more profitable comes along.

Exterior load-bearing walls

Exterior load-bearing walls are generally ordinary construction or CMU. These buildings rarely have basements, but are commonly built on poured concrete slabs. Some strip malls have brick veneer walls tied to the CMU. These combination walls are normally on the viewable sides of the building. Those sides normally out of view will be of CMU only.

A few strip malls have exterior load-bearing walls constructed of tilt-up concrete. They are constructed on concrete slabs with concrete roof assemblies. These buildings are tied together with steel rebar. Rebar is only as strong as its steel, which will begin to fail at 800°F in 5 minutes.

Even fewer strip malls have wooden exterior load-bearing walls. Of those that do, they are most likely on concrete slabs. These wooden buildings are made of the smallest materials possible that still meet code and hold up the structure. Early collapse should be anticipated. If the owner doesn't care, why should you?

Roof assemblies

Roof assemblies vary from very sturdy to so flimsy that walking on them, even without the effects of fire, gives a spongy sensation. Several roof assembly types are common. Each has its advantages and disadvantages.

If you are unsure as to the composition of the roof assembly of any strip mall, act as though it is being held up by trusses. Wood truss and steel bar joist are probably the two most common roof assemblies with wood truss being the more frequent. Wood trusses lack mass. They are constructed of the smallest material possible to hold up the intended load. Additionally, trusses are held together by metals that fail at less than 800°F in 5 minutes.

Several types of wood truss roof assemblies are in use today. Flat roofs utilize parallel wood trusses or plywood I-beams as rafters. Plywood or particleboard sheathing is then placed on the rafters and a weatherproof built-up covering applied. Roof assemblies that have a pitched-roof appearance are generally made of triangular "fink" wood trusses.

Occasionally, ordinary wooden 2×6-in. wood is used, but they can employ up to 2×12-in. rafters. However, strip malls with real wooden rafters are generally older and rarer. Pitched-roofs on strip malls are commonly finished with asphalt shingles.

Steel bar-joists are used on some flat roof assemblies. These are covered with Q-decking and a weather proof covering normally of tar and gravel built up roof covering.

Regardless of the shape of the roof, treat all roof assemblies on strip malls as though they are held up by trusses unless you're positive they're not. If the area, either cockloft (flat) or truss loft (pitched), is involved in high heat or fire, let the roof vent itself. With the size of the materials used, holes may be poked and then opened up with pike poles from aerial platforms. However, this is a slow and tiring process. The roof will open up on its own soon.

Of all the assemblies discussed above, only the true rafter-supported assemblies can be trusted for topside ventilation. These rarely have general and precipitous collapse. They sag and become spongy when weakened by fire.

Floor assemblies

Floor assemblies for most one-story strip malls are poured concrete. Poke-throughs and utility raceways will be the only general concern for fire spread. Those strip malls that have basements will usually have wood floor assemblies. Older buildings will have normal "2-by" floor joists. 2×6-in. or 2×8-in. joists are the norm. Depending on the span, these may or may not have columns or girders for additional support. Some floors may be supported by load-bearing block walls in the basement.

Most newer two-story strip malls have wooden truss or I-beam joists that support the second floor. Older malls will have true "2-by" solid joists. If the mall is built on a concrete slab, then the ceiling is the only area where fire spread will be abundant.

Prioritizing Search

To be blunt, search is not a high priority at a strip mall fire. In almost every case, occupants and patrons in a strip malls are expected to be awake and alert. These are businesses that are open during daytime and early evening hours. They close from 9:00 p.m. and don't re-open until the next morning. These are not occupancies where people are expected to be in bed asleep. This provides early notification in the event of an incipient fire.

There are exceptions to almost every rule. There are physician and dental occupancies in strip malls that perform medical procedures can put people in a drowsy state. In these occupancies, there should be personnel working who are available to assist in removing patients. At the very least they can notify firefighters of the number and location of patients immediately upon our arrival.

There are also eating and drinking establishments located in strip malls. We all are aware of the negative effect that overindulgence in alcohol can have on one's ability to maintain a high level of awareness. Alcohol intoxication could prohibit early egress in the event of an incipient fire. For the most part, I would expect any viable occupants involved in a fire in a strip mall to self-evacuate prior to the arrival of fire personnel.

Searches are required in these occupancies if they can be done safely. It should be evident that these buildings are extremely susceptible to early collapse. Many strip malls are constructed with steel roof assemblies supported by materials that will begin to fail at around 800°F in 5 minutes. That's not a very hot ceiling temperature. Since we expect most strip malls to be empty, you should proceed directly to attack evolutions unless you have specific information about possible victims. If fire conditions are small and knocked down early with minimal exposure to unprotected steel, then searches can be conducted as soon as visibility is between 5 and 10 ft.

If it is suspected that civilians are still in the building and in danger, then rapid searches should be conducted and completed ASAP. This is an instance where primary searches should be done immediately. Secondary searches should be conducted only after the building is vented, visibility is at least ceiling height, and a cursory inspection of the structural stability of the roof assembly has been conducted.

If you suspect a working fire has affected the stability of the roof assembly, but you also believe that deceased civilians are still within the occupancy, then call in heavy equipment to remove the roof assembly as carefully as possible prior to conducting the secondary search. This maybe a very unpleasant decision, but let me ask you this: is the life of a brother or sister worth the act of retrieving an obviously deceased civilian?

I didn't think so.

Searching Strip Malls

If conditions indicate that a primary search needs to be conducted, then it should be conducted as rapidly as possible. Unprotected steel begins to fail within 5 minutes at 800°F. That statistic leaves little time to conduct a primary search. Routine searches should be conducted at a working fire in a strip mall only if there is a high degree of certainty that viable victims are still within the occupancy. Strip malls are no place to conduct a search simply because you have a crew available. This is especially true of a working or advanced fire.

In figure 23–1, six bays constitute this individual strip mall. As a nugget for those concerned with mastering building construction, the definition of a bay is *the space between two rows of columns.*

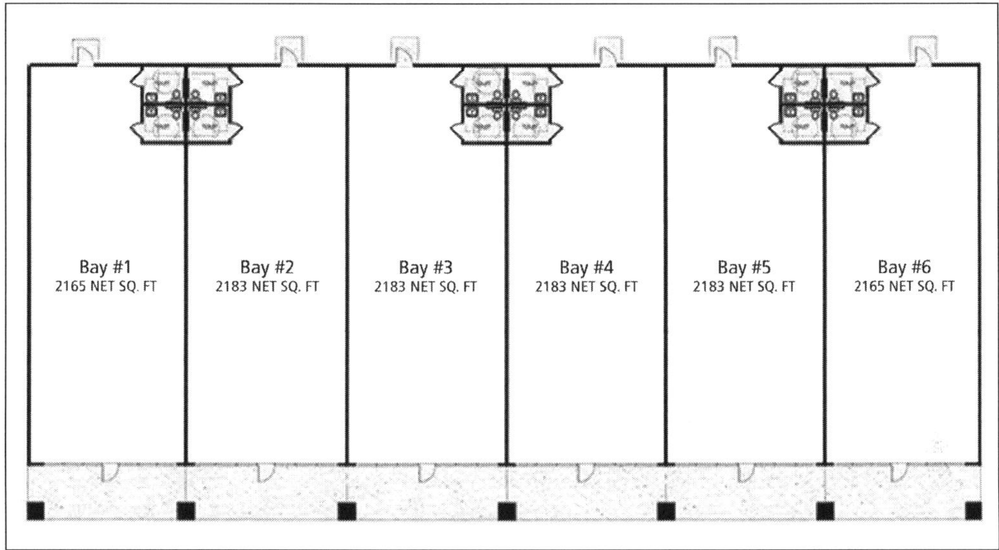

Fig. 23–1. A six-bay strip mall

For the sake of this text I will divide strip mall occupancies into two basic categories. One will be mercantile, which is an occupancy that sells some type of product. That can be anything from clothing to bicycles, prescription drugs, or food and drink. The other occupancy type is a health care facility similar to doctor, dentist, and eye care facilities we've all visited. Many other occupancy types are contained within strip malls, but to try to cover each of these individually would be too cumbersome. My suggestion is to get into every strip mall in your response district and familiarize yourself with what is contained inside. Also take note of what conditions would help or hinder a search in that occupancy.

Standard search

In conducting a standard search in a strip mall the officer would follow the attack line in through its point of entry. The occupancy type will determine what is encountered within the strip mall. For example, if it is a clothing store, the officer should anticipate encountering racks of clothing throughout the sales area with an ample storage and office area in the back of the store (fig. 23–2). If the occupancy is a dental office, the officer should anticipate a small waiting area with a partition the full width of the occupancy. Behind it will be individual examination areas and offices.

Fig. 23–2. A clothing store in a strip mall

When conducting a standard search in a mercantile occupancy, I expect the officer to remain with the crew and keep them intact as the search is being conducted. The officer should conduct a left- or right-handed search. The officer needs to maintain contact with the orientation wall while searchers move out laterally toward the center of the occupancy. Figure 23–1 shows a total of 2,183 sq. ft. in each occupancy. This equates to approximately 25×87 ft. dimensions. A crew of three including the officer would have to space themselves approximately 4 ft. apart in order to cover the occupancy in one trip—that would be down one wall and back up the opposite wall to the original point of entrance. They need to form a line approximately 12 ft. out away from the wall. A crew of four should space themselves approximately 3 ft. apart.

Health care occupancies are generally set up with a small waiting area in the front of the occupancy. Once you pass that, you will encounter individual examination areas and offices toward the rear of the occupancy. Using standard search, the officer enters the waiting area and sends one or two firefighters in one direction. The officer and any remaining firefighters search in the other direction until the waiting area has been covered.

At that time they all enter the examination areas. They are likely to contain a hallway with small examination rooms branching out from either side (fig. 23–3). This can be searched in much the same way as individual sleeping areas in a small house off of a main hallway. Upon entering a room, one or two firefighters go to the left and the remainder of the crew heads right until the room has been completed.

As you can envision, that is an awful lot of firefighters in a very small examination area. I believe standard search to be an extreme waste of time in this case. It has required me to re-evaluate my preference of the standard method of search in strip mall occupancies.

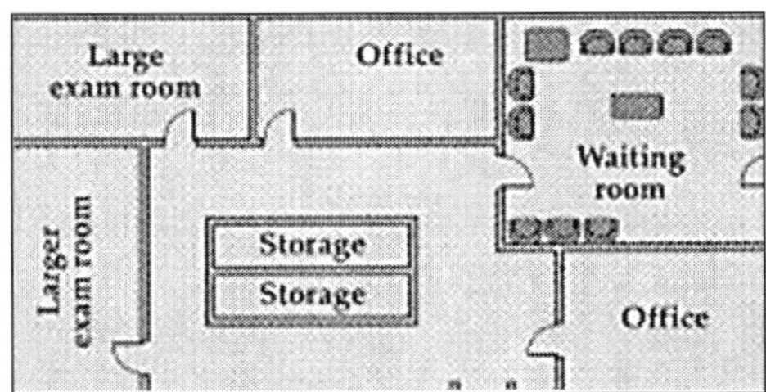

Fig. 23–3. A health care office in a strip mall

Team search

Due to the size and occupancy type of strip mall occupancies, I do not consider team search a viable option. In chapter 22 I discussed attempting team search in an eating establishment. The same rules hold true for nightclubs and drinking establishments. Unless we can somehow elevate the search rope to a height that prohibits interference from tables and chairs, then my first option would be to aggressively vent and provide myself 5 or 10 ft. of visibility. That should be a mandatory minimum if a search is required due to life safety conditions.

The same holds true for health care occupancies in strip malls. The purpose of a search rope is to maintain orientation with egress in large, wide open areas. Those conditions aren't usually present with most occupancies in strip malls. In chapter 27 I discuss big box stores, which use similar construction as strip malls but are very different from the traditional strip mall.

Oriented search

For several reasons I consider the oriented search to be the search of choice for a primary search in a strip mall.

If using the oriented method of search in the mercantile occupancy, I would pull a 2½-in. hoseline inside the building to create a high wall with which to maintain orientation to my point of egress (fig. 23–4). The hoseline should be stretched into a point where the search will begin, or until the hoseline is pulled to its fullest.

At that point the oriented officer turns and faces the point of egress. The officer stays on top of the 2½-in. hoseline while the searchers work out laterally until they meet a wall or cover a predetermined distance. Using figure 23–1, with the officer on the 2½-in. line in the middle of the occupancy, each searcher would only have to move about 12 ft. laterally away from the line. This can be accomplished by either using a preset number of side crawls or employing webbing to guide lateral movements away from the oriented member. Once the searchers move out away from the oriented officer and return, the crew moves 3 or 4 ft. closer to the point of egress, and the searchers repeat their lateral searching motion. Repeat this until the point of entry has been reached. At that time an "all clear" should be given to the incident commander. The oriented officer will make a determination about whether the search crew should exit the building or search any uncovered areas in the occupancy.

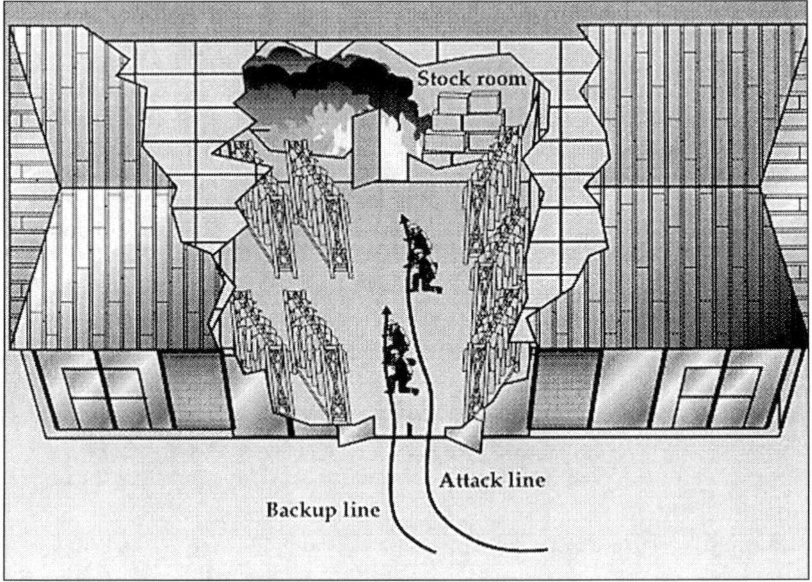

Fig. 23–4. A mercantile store in a strip mall

When conducting oriented search in a mercantile occupancy in a strip mall, search it the same as the first floor of a ranch style home. The waiting area will likely be the first area encountered, followed by individual examination rooms leading from a common hallway. To cover the waiting area, send one or two firefighters to the left and the other firefighters to the right until the two teams meet somewhere in the middle and the area has been covered. We then proceed into the individual examination areas leading off the hallway. Cover those by sending an individual searcher into each examination or office area that I encountered.

Maintaining Continuity of Search

Except in extreme circumstances such as an arson fire or other rapid catastrophic ignition, I would not expect to locate a large number of victims in a strip mall. As such, maintaining continuity of search may not be that difficult. Anytime I can get the searcher who locates the victim to pass that victim on to another firefighter and return to the exact location where the victim was located to continue the search, then I believe I have provided the best scenario to locate and remove any more yet-to-be-located victims. In my opinion, this is the epitome of an unselfish search that emulates the highest tradition of the fire service. Anything short of this does a disservice to the civilians that we serve.

Places of Worship 24

Although construction and the intended use of these occupancies are very dissimilar from strip malls, fires in houses of worship do have one thing in common with strip mall fires. They both carry a low expectation of large numbers of endangered civilians. That is because people in churches, synagogues, mosques, and other places of worship are expected to be alert and awake during their visit.

Problem Associated with Searching Places of Worship

Although I do not expect a working fire in a place of worship to produce a large number of fatalities (with the exception of an incendiary fire during worship activities), the sheer area requiring coverage provides challenges for the search officer. With the exception of converted churches (see below), these occupancies are fairly symmetrical. Search patterns should be relatively simple.

Building Construction 101

Before discussing the construction associated with worship building fires, I need to describe the three main types of worship buildings that exist today for the purpose of this discussion.

The first type is the "old traditional Gothic" church or cathedral style. These structures are usually found in the older parts of communities, and are generally made of non-fireproof materials using ordinary stone or brick construction.

The second type of worship building is "modern traditional." These were built in or around 1950 and are more fireproof in construction. Many have built-in fire detection and protection systems. Also, fire-stops in interior walls may be present. These tend to be smaller in size than the old traditional Gothic buildings. Although the architectural lines are the same, they provide an easier atmosphere for firefighters to work in.

The last type of worship building I will discuss, I will term "converted churches." These are buildings that were initially designed for a purpose other than as a place of worship. These can range from old balloon-frame storefronts to converted storage facilities to schools, gas stations, or strip malls. Although it's a challenge, I will briefly try to explain the construction types of the three categories of places of worship.

I do not intend to overlook other types of worship buildings that might have different layouts or construction types. It is incumbent upon local firefighters to become familiar with the worship center structures in their neighborhoods so that they will be prepared.

Exterior load-bearing walls

Exterior load-bearing walls of old traditional Gothic churches are predominantly made of stone or brick, mortared together. In other words, they employ true ordinary construction. These walls are stiffened by placing piers or buttresses throughout. The height of these walls can range from 20 ft. to over 50 ft. As with any ordinary constructed wall, the higher the wall, the thicker the base (or pier or buttress). The interior of these walls can be solid stone, brick, or can have a wood, non-load-bearing wall with lath and plaster to provide a less cold finish. The latter will provide an abundance of fuel if fire enters the area. The channel of the fuel leads to the roofline, where more fuel is present.

Modern traditional worship buildings have the same general construction feature present with their older counterparts, but are usually built with more safety and fire-stopping measures in mind. Exterior load-bearing walls will be made predominantly of stone or brick. Steel will be more prevalent as wall supports, piers, and buttresses are more decorative than functional.

Almost any type of exterior load-bearing wall can be present in converted churches. Everything from lightweight steel to balloon frame to more traditional ordinary construction can exist. Know your buildings.

Roof assemblies

Roof assemblies on old traditional Gothic buildings can be hard to classify. Most are peaked roofs of great pitch. Exterior coverings are generally slate or terra-cotta tile. However, some metal coverings can be found. These roofs are applied to wooden planks or large purlins. Roofs are supported by massive timbers (rafters) using simple truss frames. Some are arched and supported by hammer beams (short horizontal members attached to a principal rafter). These rafters can support spans of over 20 ft. They are supported on exterior walls by corbels or imposts that are tied to piers in the exterior load-bearing wall.

Roofs of modern traditional worship buildings look much like older types but tend to be less massive. Mixtures of architecture styles are found in these structures. Exterior roof coverings can be of slate, terra-cotta tile, metal, or wood shingle. They are laid over wood plank, purlins, or plywood. These roof assemblies are supported by solid or laminated timbers as rafters. The spans of these massive rafters can be very wide. They are tied to walls on steel columns with decorative wood or plaster trim that resemble corbels or imposts. You must remember that steel maybe present in exterior walls to support loads. Some modern worship buildings have very ornate and geometric designed rooflines. Suspect truss assemblies and all the problems associated with trusses. Be very skeptical of roof ventilation in unfamiliar shaped/sloped rooflines.

Roof assemblies in converted churches are constructed of a variety of materials. Wooden flat roofs with cocklofts can be found in converted taxpayer storefront worship centers. Gas stations converted to churches can have truss roof assemblies or wood. Modern strip malls will likely have truss roof assemblies.

Floor assemblies

Floor assemblies in old traditional Gothic churches are usually constructed of wood planks with a myriad of coverings. Churches in use today may have carpet, vinyl, ceramic tile, concrete, or the original wood floor. Floors will be supported by large wooden joists. These joists are supported by girders and/or columns, commonly of stone or masonry.

Modern traditional places of worship often have poured concrete floors supported by columns or slab floors. Some will have wooden floors with wooden floor joist systems. That is usually only seen in smaller worship centers.

Converted churches can have a myriad of floor systems. It's best to just get inside and check. Even before you arrive, you should know your buildings!

Prioritizing search

Because visitors to places of worship are in a more alert state than in other occupancy types, the chances of having many victims in a working fire are low. To this is not to say a fire at a place of worship has not caused significant civilian fatality in injuries in the past, but in recent history I cannot recall such an incident where significant civilian fatalities occurred. The only exceptions are the few caused by arson and accelerants.

Searching Places of Worship

Before discussing search specifics, it is imperative to provide one fire attack rule that I believe significant in fighting any fire in these buildings. I can see no reason why the first hoseline in one of these fires should not be a 2½-in. or greater attack line. A fire in a house of worship is no place to play catch-up. Come in with adequate water or don't come in at all.

Standard search

In attempting to conduct a standard search in an older traditional place of worship (fig. 24–1), the officer and crew enter the building at a location that provides the most rapid access to the area where victims are believed to be. Because there is often fixed seating in these occupancies, searches are somewhat easy to accomplish. Church and synagogue pews are generally fastened to the floor in some manner and do not move easily. They can be used as a point of orientation.

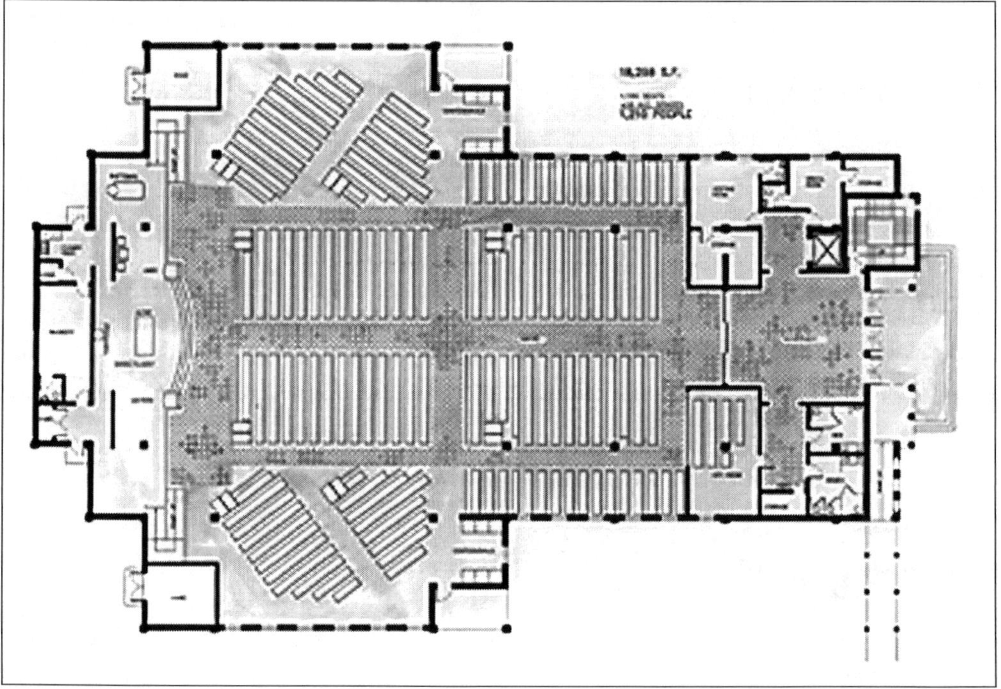

Fig. 24–1. Traditional place of worship layout

If the information gained prior to entry leads the officer to believe the search should be conducted mainly in the sanctuary area, then the officer should proceed to that area and commence a search. It would be best to start at one end of the sanctuary where it meets the congregation seating area, and then work laterally across the face to the opposite end of the sanctuary area. Normally there is a set of steps that lead from the seating to the sanctuary area. That means the pulpit is usually elevated at least four feet above the congregation. The officer maintains contact with the wall along the sanctuary area. The officer instructs the searchers to move out laterally toward the congregation seating area, then back to the officer. As a team they move from one end of the sanctuary to the other.

If searching the congregation seating area using the standard method of search, the officer can keep the crew intact or split into two teams if staffing allows. If you split, have each individual team search a set of pews until the congregation seating area has been covered.

In the case of a church, if the narthex contains more areas such as confessionals and baptismal fonts, they need to be searched. Enter through the main entry doors near the narthex and either do a left- or right-handed search. The officer follows the wall of orientation and instructs the searchers work out laterally until they reached the congregation seating area. Then they move back to the officers location and proceed around the entire narthex in this fashion.

If attempting to search a modern place of worship using the standard method, I would follow a pattern similar to that described above. Proceed to the area where victims are believed to be, then systematically cover that entire area using the principles of a standard search. You must keep the crew intact in each individual area prior to moving on to the next. In some instances, the crew can be split into two teams and two searchers head to the left and the other two go right. When they meet, that area is complete.

Figure 24–2 shows a modern church. There are many small areas in the rear of the church opposite the sanctuary or altar. There are classrooms, chapels, kitchens, offices and other areas that will make any search in these occupancy types a challenge. This may be an instance where a 2½-in. hoseline is stretched into the area to provide a means of orientation with the path of egress.

Because of the vast number of buildings that have been converted into churches is far beyond the school of this text and my intent to cover each and every possible scenario. It is important that individual crews inspect and pre-plan these occupancies to determine the best way to conduct a search in zero visibility.

Team search

Team search works well to provide a point of orientation in large, wide open areas such as warehouses and big box stores. For older traditional and modern sanctuaries, fixed seating provides the best point of orientation. The nave and other aisles in these sanctuaries provide an easy means of identifying the general area of egress. I do not believe that team search provides a better means of search in these facilities. I am aware that many worship centers have gymnasiums and other large open areas attached to them. These areas are much better laid out to utilize team search.

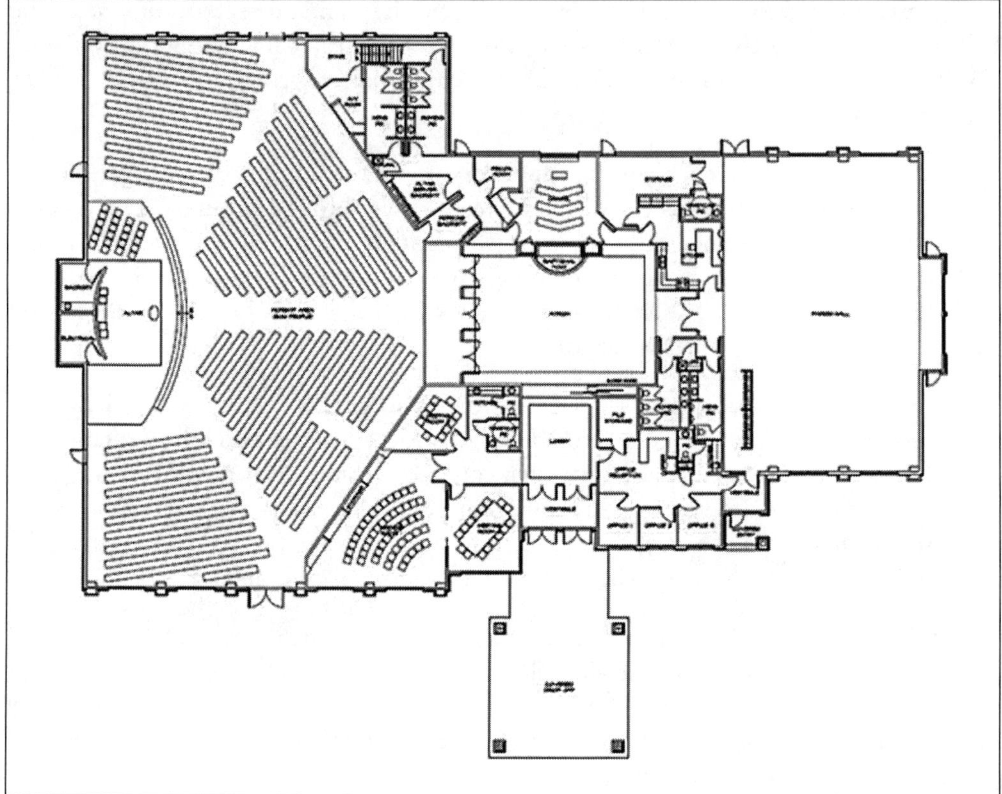

Fig. 24–2. Modern place of worship layout

Oriented search

The oriented search works very well in these occupancy types (fig. 24–3), with the exception of the converted church, which I will discuss later.

The oriented officer is positioned in the middle aisle in the sanctuary. The oriented officer then sends the searchers laterally down each set of pews. The searcher can run a hand along the seat of the pew to cover that area and occasionally sweep under the pew with the other hand to check for victims there. This tactic has proven effective in traditional and more modern worship building types.

When searching the sanctuary area, the officer maintains contact with the wall, similar to the standard method of search. The searchers work out laterally until they meet in the congregation seating area.

Due to the vast number of different types of converted churches, it would be unrealistic to try to cover all of them. It would simply take too much time and probably still miss something that is present in your jurisdiction. Get out of the station to visit and pre-plan in your district.

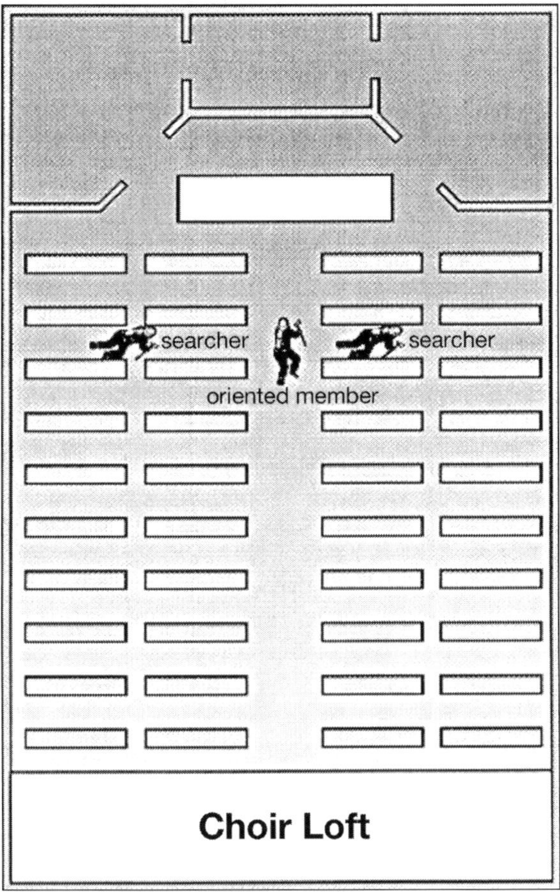

Fig. 24–3. Oriented search is an effective way to search church or synagogue pews.

Maintaining Continuity of Search

Continuity of search is not a huge concern in worship buildings because it is not expected to find many victims in those fires. Early notification by the victims and their state of alertness at the time of the fire should keep victims to a minimum. As with any search, it is vital that once a victim is removed, the searcher return to that point. Ideally, the searcher who originally located the victim should continue the search. If is not possible, it is incumbent upon the officer to glean information from the searcher who located the victim in order to maintain continuity once another searcher is available. Anything that can be done to stop duplication of effort or missing areas entirely will save lives in the long run.

Industrial Occupancies 25

Industrial occupancies pose the same concerns as converted churches with regard to the huge number of possible construction types and layouts. When I discuss industrial occupancies, I envision any building where products are made, altered, constructed, or otherwise fabricated. Some of these facilities can be as large as automobile plants while others are only slightly bigger than a large residential garage.

Problems Associated with Searching Industrial Fires

These occupancies are places of business and occupants should be expected to be in an alert state. However, the volatility of the processes and materials used on-site can cause devastation and serious fires. Much consideration should be given to commencing any search unless it can be conducted on our terms. It would be impossible to explain to a widow or widower why a firefighter was killed while saving unassembled or broken industrial products. Risking firefighter lives to save victims is one thing; risking them to save a broken down Chevy is another.

Building Construction 101

Construction types will vary. However, some generalities can be made. Many will be of fire resistive construction. Ordinary (CMU in newer and brick in older) and concrete construction are used in many of the manufacturing, industrial, and storage occupancies around the country. Some steel building, and/or a combination of ordinary, concrete and steel construction is used. These buildings are meant to be functional and not pretty. In newer communities or those with tightened building codes, some of can be both functional and "pretty." Please remember that behind the façade, the same dangers lurk.

Exterior load-bearing walls

Exterior load-bearing walls will be predominately ordinary, concrete, or steel building construction as mentioned above.

Roof assemblies

Roof assemblies will be mainly steel bar joist, supported by a beam and girder system. On top of the bar-joists will normally be Q-decking with a metal deck or built up roof. The roof usually has a tar-and-gravel or membrane covering. Some smaller buildings may have plywood decking. Some will also have poured concrete roofs on Q-decking or an all-concrete roof assembly.

Floor assemblies

Floor assemblies are generally concrete. Some will have bar-joist with beam and girders but concrete will be the majority of floors. Some older buildings may have wood plank floors. Many of these will have years of spilled oil and other combustible materials soaked in. Fire can spread with unbelievable speed on these floors. Know your buildings.

Prioritizing Search for Industrial Fires

Except for flash fires that could trap workers, these occupancies generally do not produce large loss of life fires. That is not to say that there has never been a fire in an industrial facility that produced a large number of fatalities. Because these are occupancies where people go to work and not sleep or rest, most workers will be in an alert state at the inception of the fire. That allows for early warning and egress from the facility.

Searching Industrial Fires

It's tough to discuss a thorough way to search industrial occupancies because of the vast number of different industrial occupancies throughout the United States. I will use a couple of smaller industrial floor plans in order to illustrate effective evolutions in these types of searches. In chapter 27 I discuss searching malls and big box stores.

Standard Search

The facility in figure 25–1 is one building containing two separate occupancies divided by a party wall. A party wall is a wall that separates two addresses in a single structure. One facility is just shy of 14,000 sq. ft. and the other is 20,648 sq. ft. Due to the sheer size of these two occupancies, I do not believe that a search for civilians is feasible in either.

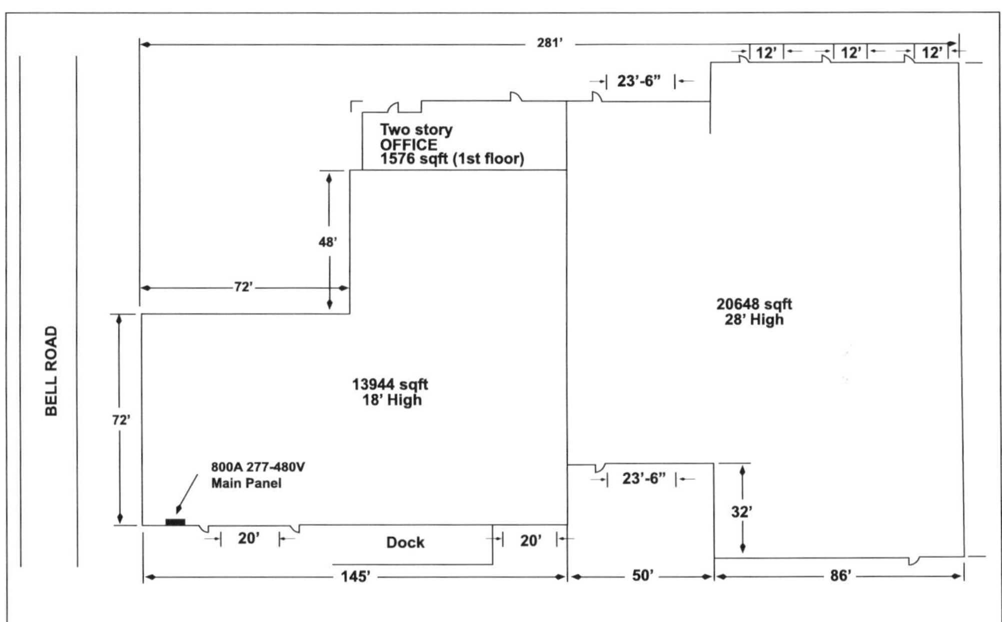

Fig. 25–1. Floor plan of a large industrial space containing two separate occupancies divided by a party wall

In order to cover this smaller facility in 15 minutes or less, each searcher of a four-person crew including the officer would have to search an area equal to a large single family home. That's approximately 3,500 sq. ft. each. Even if that were feasible, the sheer distance that would have to be covered would prohibit effective communication between the officer and the searchers.

One last consideration in searching these occupancies is what materials are contained inside. Check if the space is relatively wide open and free of material, or fully occupied with machinery, shelving, tools, and other materials. If it's full of materials, navigating in zero visibility becomes extremely difficult.

Team Search

Team search was developed specifically for industrial types of occupancies. Team search is used, for the most part, in locating downed firefighters in large, wide open areas such as warehouses.

In conducting a team search in either of the occupancies in figure 25–1, the officer must pick a point of entry that provides the closest access to the location of victims. In this case, we will use one of the doors located on the A side of the building near the dock.

An anchor point should be selected outside the building. Then the crew enters and begins to play out the search rope. Once inside, if a thermal imager is available, the officer would scan the building both side to side and up and down. The officer should pass the imager to the other members of the team so they can quickly scan the area as well. After the scan, the officer chooses a direction for the anchor to start and they proceed in that direction. The anchor plays out the rope, followed by the officer and the remainder of the crew.

In searching the smaller facility in figure 25–1, I would conduct the search with one of two methods:

- Upon entering the building and scanning and locating the wall directly ahead, the officer directs the anchor to move ahead until the wall is encountered. At that time the anchor should remain near the wall. The officer then works back toward the point of entry and instructs the searchers to work out laterally off the search line. The distance is approximately 72 ft. That would require each searcher to cover approximately 35 ft. out from the search line. This could be done safely with smaller 30-ft. tag lines or webbing, if available. If not, the officer may have the searchers crawl laterally 10 to 15 side crawls out from the search line (fig. 25–2). The officer would have the anchor move towards the officer's location as they proceed back toward the point of entry. This procedure will only cover one third of the total square footage in the smaller section of the building. In order to attempt to cover the entire building, a second method should be attempted.

- The officer would scan the building with the imager, then have the anchor proceed diagonally to the right rear corner of the facility. Once the anchor reaches that corner, a similar procedure as described above would be conducted to move back toward the point of entry.

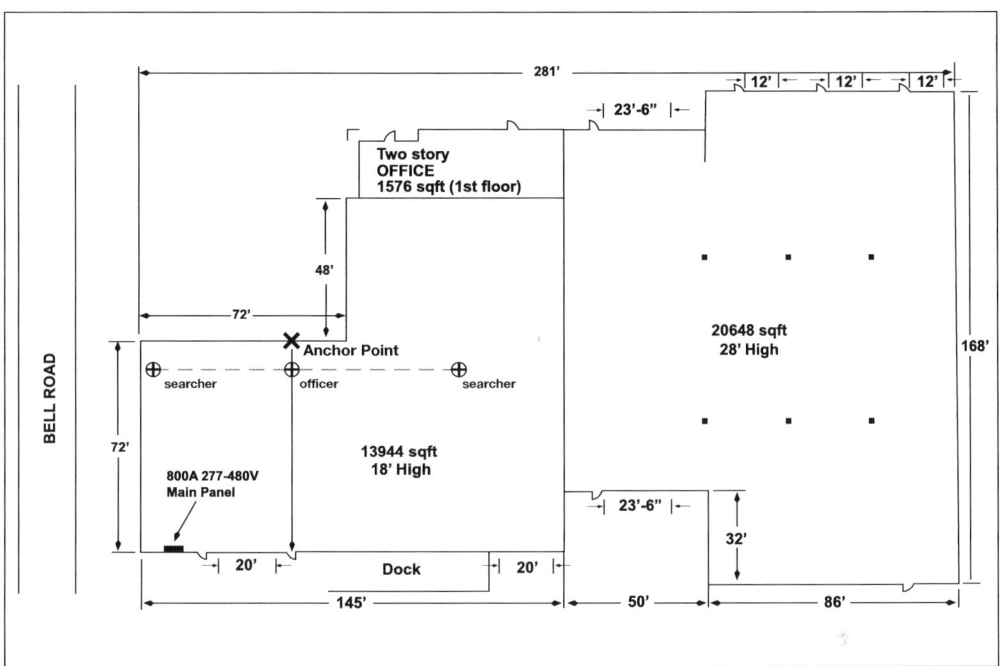

Fig. 25–2. Searchers working laterally off of search line

In searching the larger facility, there are six columns located within. These can be used to tie off the search line. Then smaller areas can be covered more easily.

The Oriented Search

In conducting an oriented search in the facility in figure 25–1, play out a 2½-in. hoseline in lieu of a search line as described above. This will provide a 2½-in.-high wall with which to maintain orientation to the point of entry. Once the hoseline is played out, it should be charged. Then the searchers should move out laterally from the oriented officer using webbing or individual tag lines, or instructing the searchers to crawl a predetermined number of side crawls laterally off that line as illustrated in figure 25–3.

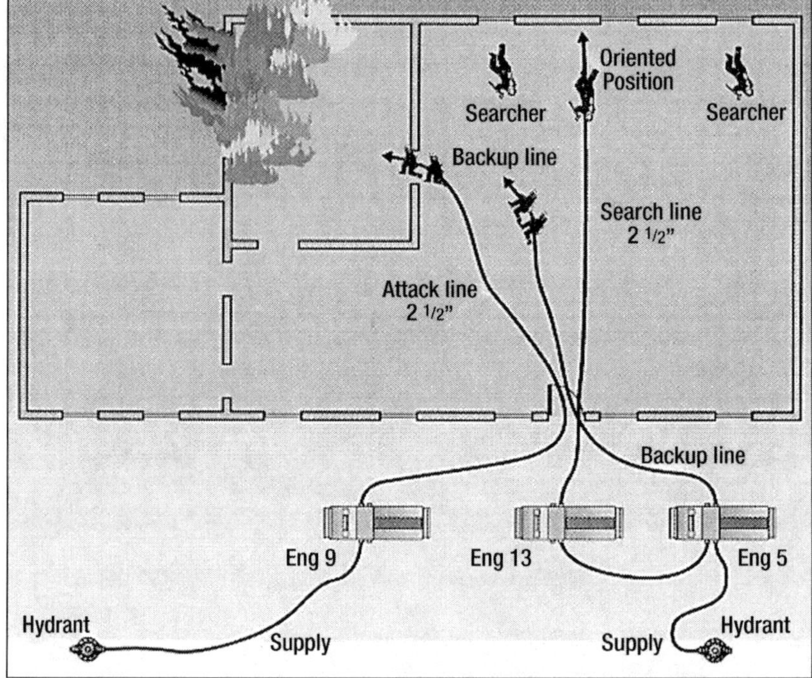

Fig. 25–3. An industrial layout with attack, search, and backup teams in place

Maintaining Continuity of Search

Ideally, early warning systems for workers in the facility will lessen the need for searches. Due to either the extremely complicated layout of an industrial plant or the openness and lack of items with which to orient yourself, maintaining continuity of search is an extremely challenging task in an industrial occupancy. A rescue group is a must. That way the located victim is removed by the rescue group and the searcher is allowed to continue on with the search once the victim has been passed to the rescue group.

Final Thoughts

Due to the sheer size, complicated layout (whether because of a large number of objects or the wide open spaces), and the construction features usually present in these facilities, I would be very hesitant to conduct an initial search in an industrial occupancy if I had zero visibility.

My thought process would be to:

- Rapidly determine the feasibility of firefighters safely entering the building and the feasibility that savable victims could still be in the building.

- As safely and rapidly as possible darken and extinguish the fire.

- Aggressively ventilate utilizing PPV, if available. Vertical ventilation is a last resort if the roof can be accessed and worked on safely.

- After the IDLH has been removed from the area, then we can go in and look for victims.

Do not attempt to fight these fires on the fire's terms unless absolutely necessary to save a life. Even then a diligent fire commander must weigh the department's ability to cover these expansive areas against the possibility of locating a victim in enough time to make a difference. If you can't do it in 15 minutes or less, then go to plan B: remove the IDLH from the victim.

Newer Office Buildings 26

Another new phenomenon that has occurred over the past 20 years or so is the rise of newer, lightweight constructed, low-rise and single-story office buildings (fig. 26–1). Similar to strip malls, these are disposable buildings made of lightweight, non-fire-resistive construction. The buildings are prone to early collapse. Layouts and floor plans can be complicated and change often.

Fig. 26–1. A new low-rise office building

Building Construction 101

There are three major groups of new office building construction types. Wood (or truss) frame, concrete, and steel are the three major types associated newer office buildings. You must recognize what type is used because each will react differently if involved in fire.

Exterior load-bearing walls

Exterior load-bearing walls in many of these occupancies are made of wood. Most of these are one- or two-story platform construction. The exterior finish or siding on these can be made of a variety of materials. Brick veneer, aluminum siding, shake siding, and combinations of any of the three are the standard.

New office buildings may be wide open with a system of columns (wood or steel with a decorative cover) or the large span may be supported by nothing but the truss assembly above. These buildings, with large spans with no interior walls to support what's above, have the potential to be the next decade's firefighter killers. Truss assemblies will support the large openings. If left covered and unnoticed, they can surprise crews with sudden collapse.

Interior wall assemblies are good. They provide protection to crews below. Roof collapses can be slowed by a system of interior partition walls. It doesn't matter if the wall is a bearing or non-bearing wall, it will lessen the effects of a collapse. However, engineers and interior designers have been directed by the building owners to create office buildings with wide open spaces filled with modular work-stations.

Please watch buildings go up in your community. On the way back from a run, take alternate routes through your district. Many departments have relationships with the building officials in their community and are notified when plans with trusses are approved. However, if you are crawling into a building and notice an uninvolved area with no interior walls, inform command and anticipate collapse if fire enters the area above. These interior wall assemblies when present can be wood or steel studs.

Another construction type used for modern office buildings will have exterior load-bearing walls constructed of tilt-up concrete. These exterior walls are pre-cast (either on-site or off) and then lifted into place by a crane. The sections are tied together by rebar. This rebar can be exposed to heat in places. It will begin to fail at 800°F. Collapse can occur, usually as whole sections of wall are pushed out. This will also make roof assemblies unstable. Most of these roof assemblies are held in place by little more than gravity. These buildings often have steel stud interior partition wall systems. These walls are there to divide floor space and do not add to the structural stability.

The last type of exterior load-bearing wall used in the newer office buildings is steel assemblies. They are unprotected steel skeletons that support exterior glass curtain or panel walls. Interior wall assemblies can be wood, but the majority will be steel stud systems. Interior steel columns are protected by encasement (drywall) or masonry.

Roof assemblies

Roof assemblies in wood frame buildings are usually supported by truss assemblies. Some of these buildings have triangular truss replicating a gable roof while others form different roof configurations. The trusses have plywood or cheap chipboard sheathing and asphalt shingles.

Newer office buildings that have glass panel or curtain walls usually have steel bar joists supporting the roof. Q-decking used as sheathing or built-up roofs are the norm. Tilt-up concrete buildings normally have concrete roof assemblies.

Floor assemblies

When a new office building has wood platform construction, floor assemblies are made of truss or wooden I-beam construction. Both of these are bad, especially if there are large spans below. Collapse can be a real concern with these assemblies. Plywood or cheaper particle board is placed over the wood truss assembly. Carpeting is the normal outer floor surface. Linoleum and simulated hardwood floors can also be found.

In those steel and or concrete buildings with panel or curtain walls, look for steel bar joist floor assemblies. The first floor can be a concrete slab. A newer, cast-in place concrete floor system is being used in my area. Concrete is poured onto thin steel panels, similar to Q-decking but with smaller grooves. In the past, plywood forms were used. Then the plywood was removed after the concrete cured. In this new process, they leave the metal decking in place after the pour. Regardless, these floors are supported by steel bar joists and will collapse at 800°F.

Prioritizing Search in Newer Office Buildings

Search will not be a high priority in these occupancies, for the most part. As with the last few chapters, early notification and victims' state of alertness all lead to rapid egress in the event of an incipient fire. Some of these occupancies, such as doctor and dental offices, do indeed have victims that may be in a lowered state of alertness. However, I would expect that nurses and other staff would assist an early evacuation, or at the least be able to inform first-arriving firefighters that there are victims within the occupancy.

Searching Newer Office Buildings

Floor plans and layouts in these buildings can range from very simple to quite complicated, depending upon the occupancy used for the specific building. Figure 26–2 shows the layout of an office building used as dentist and physician offices.

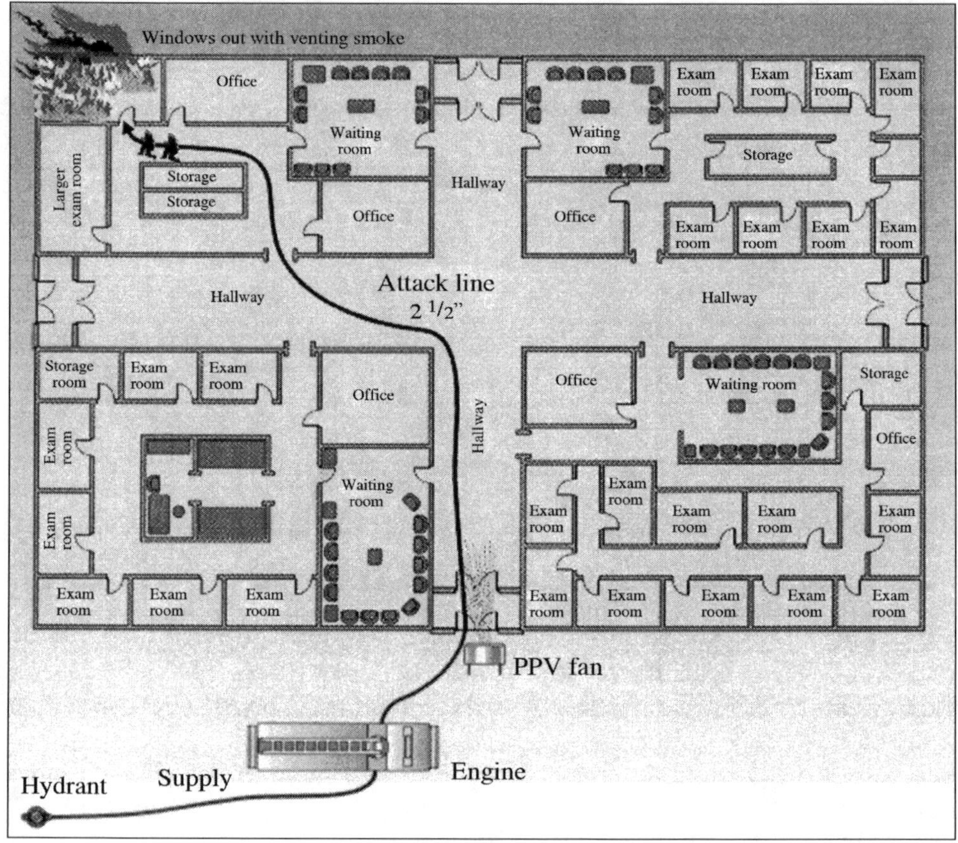

Fig. 26–2. The floor plan of a doctor or dentist office

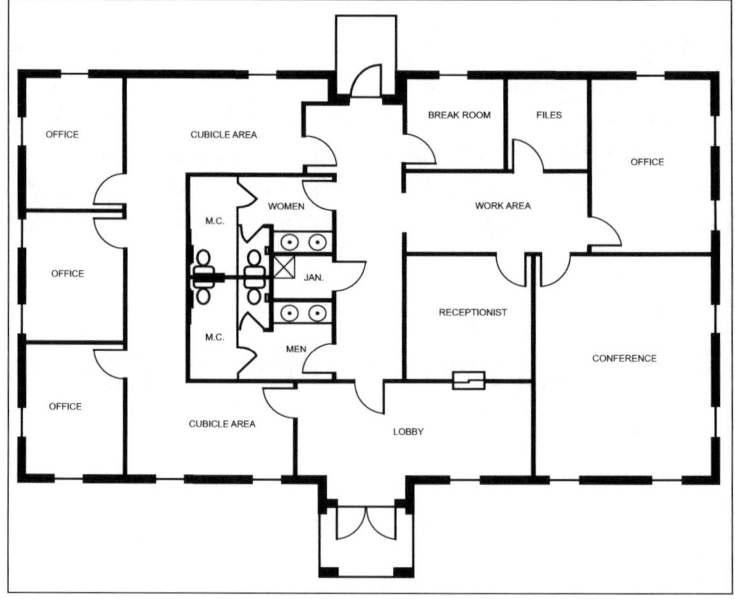

Fig. 26–3. A law or accounting firm office

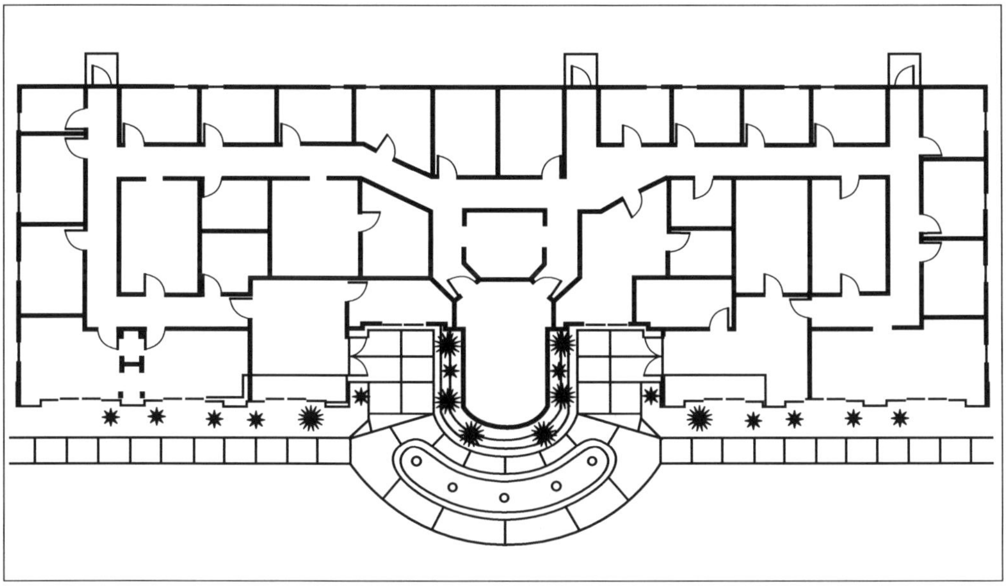

Fig. 26–4. A large facility with many small offices

Figure 26–3 shows the floor plan of a building occupied by one company and used as business offices. This could be an accounting firm or lawyer's office. Figure 26–4 is a very complicated facility that could be many small individual offices or one large company with many offices. The size of these buildings can vary from 1,000 sq. ft. to tens of thousands of square feet. The standard rule is that any occupancy less than 12,000 sq. ft. is not required to be sprinklered. Buildings larger than 12,000 sq. ft. must be protected (sprinklered) or subdivided by a fire rated separation. However, older office buildings lacking sprinkler protection with larger areas exist in many areas and have been grandfathered, meaning that they are not required to meet current building code requirements.

Standard Search

Newer office buildings can be searched the same way we search a large, ranch-style home. In figure 26–2, the officer starts the search as close to the fire as possible where savable people could be, then works out from that point. Say there is fire in the BC quadrant of figure 26–2. The officer should take the crew into the occupancy and check with the attack officer to ascertain fire conditions in the area. Let's assume for this scenario that the attack officer believes that no savable victims could be in the involved office. At that time the officer should take the crew to an adjoining office in the hallway and begin search there.

Assuming the search starts in the office in the AB quadrant, the officer would either begin a left- or right-handed search. If using a four-person crew, two members go to the left and the officer and a searcher go to the right and begin to work the hallway. Upon encountering an individual room, the search team enters. One searcher goes to the left the other heads right. They inevitably meet somewhere in the middle. At that time they move back out of the room and proceed down the hallway to the next one.

If you're using a three-person crew, each room is searched by the entire team. One firefighter goes one direction and the other two moves in the other. Once they meet in the middle, all three proceed out of the room and back to the hallway.

Figures 26–3 and 26–4 should be searched in a similar fashion. The officer determines the oriented wall and counts doors as the team proceeds through the building in order to assure orientation with the point of entry.

Team Search

Due to the size of the building and the smaller individual rooms contained therein, team search is not a viable option in this circumstance.

Oriented Search

The oriented method of search provides a safe, systematic method of searching newer office buildings. The officer can focus on the location of the crew as it relates to the point of orientation as the search progresses. The officer sends the searchers to enter the smaller individual office areas to conduct one-person searches, using the hallway as the point of orientation to the main egress. Once inside a specific occupancy, the officer would use smaller individual hallways for orientation

Figure 26–2 contains individual health care offices. The fire occurs in the office in the BC quadrant at the left rear of the building. The oriented officer proceeds to that office and ascertains if conditions are suitable to survival for civilians. This information usually comes from the attack officer. Assuming conditions were not conducive to life, the oriented officer takes the crew to an adjoining office in the hallway to begin search. Once inside the individual office, the oriented officer should lead the crew around the interior of that office.

Assume the officer begins search in the AB quadrant of the building. If doing a left-hand search, the officer would encounter an office and have searchers begin search there. The officer proceeds down the hall to the waiting room and instructs another firefighter to begin search in that area.

Understanding the typical layout in a modern office buildings such as this, the prudent fire officer understands that the possibility exists that there could be offices or rooms off both sides of the hallway. Understanding that, the officer should maintain the wall on the left as the wall of orientation. However, the officer should "ping pong" back and forth down the hallway, moving forward looking for doors. That way nothing will be missed.

Each firefighter informs the oriented officer of the search direction and number of walls and rooms, then starts a search. If the officer has a third searcher, the next room should be located and that firefighter should start search there, as well. At that time, the oriented officer should return to the other two searchers to check their progress and safety. This process should continue until the entire doctor's office has been completed. Upon completion, the oriented officer gives an "all clear" in the AB quadrant and proceeds to one of the other two remaining offices. Then search continues in a similar manner.

The other two occupancy types illustrated in figures 26–3 and 26–4 should be searched in the same manner, with the exception that the entire office building can be searched off one main hallway. If doing a left-handed search, the officer should keep the hallway wall to the left as the oriented wall. The officer progresses down that wall, searching both sides of the hallway for individual rooms. The search officer must move around the entire interior hallway, allowing the searchers to check the smaller rooms that are located.

It is imperative for the oriented officer to count doors while moving throughout the hallway. This is necessary in the event that a rapid egress is required. The officer should also remind the searchers to check for the presence of windows as they search individual rooms. The officer shouldn't direct the searchers to open or break windows as they are encountered. Instead, remember their location in the event that alternate egress might be required.

Maintaining Continuity of Search

Maintaining continuity of search with multiple victims could become a challenge in these occupancy types. It must be remembered that business office buildings should be occupied by individuals in an alert state of awareness, whereas health care occupancies may have victims in a less responsive state.

If the potential exists for multiple victims, a rescue group becomes a necessity in newer office buildings. The oriented officer should do everything possible to assure a new searcher completes the room in which a victim was located. An even better option is to get the original search team back into the room to continue the search. There is no better way to maintain continuity of search.

Large Malls and Big Box Stores

27

The big box store concept has become a more prevalent occupancy type in the last 20 years. These buildings are generally retail superstores that sell a wide range of merchandise, from tools, building supplies, and large household appliances to groceries and pharmaceuticals. Enclosed malls are not quite as new and have been around for 40 and 50 years. Most everyone should be familiar with these structural types.

Problems with Searching Big Box Stores and Malls

Lightweight construction heads the problems list with searching big box stores and malls. These buildings are constructed similarly to strip malls. They are disposable buildings, not constructed to resist the effects of fire. Due to their size, they are required to be fully sprinklered if they were built after 1975.

Another concern is crew ability to search a structure of this size successfully and in a safe amount of time.

Building Construction 101

In general, big box stores are constructed of tilt-up concrete construction. That consists of steel construction utilizing concrete masonry units (CMU) as the exterior walls of the building. Except where covered by code, I would expect the steel that is providing the load-bearing capacity of the rough assembly to be unprotected. That makes them extremely vulnerable to early collapse.

The good news is that these buildings are generally much larger than 12,000 sq. ft., and therefore required to be fully protected by automatic sprinkler systems of sufficient volume to quickly handle incipient fires. However, this is true only if the design and the water supply are adequate. Changes in commodities could increase sprinkler demand that can and have overtaxed sprinkler systems. Inadequate sprinkler systems can result in total collapse in 10 minutes or less.

You will normally find two types of enclosed malls: newer lightweight constructed malls or old refurbished buildings that have been converted into an enclosed mall. Each type has specific construction concerns. It is necessary to review each separately.

Exterior load-bearing walls

Exterior load-bearing walls in big box stores are usually tilt-up, precast concrete construction. These buildings are essentially held together by steel cable, which can fail at 800°F. Steel exterior load-bearing walls can have several sidings attached to the steel columns. Steel siding, brick, aluminum, and even wood can be applied to the steel walls. Concrete slabs similar to tilt-up construction can also serve as a siding. The other main method of constructing the exterior load-bearing walls in big box stores is to construct a skeleton with columns of steel. They support the structure and its roof assembly. Then the exterior weatherproof wall will generally be constructed of CMU.

On the older converted mall, exterior load-bearing walls were primarily constructed of brick, block, or a combination of steel and brick. With ordinary constructed buildings, the interior columns will be of ordinary construction, steel, or a combination of the two. Some store fronts of old taxpayers have masonry exterior load-bearing walls and simple wood joists that span from party wall to party wall. Some of them have been converted to larger, enclosed malls. The original party walls that once separated occupancies have been breached to allow for the interior movement of shoppers, all within the confines of four now rather large exterior walls.

In newer lightweight malls, exterior load-bearing walls are of steel construction, concrete, or a combination of the two. These buildings are fire-resistive and should by code be protected with automatic sprinklers if their square footage exceeds 12,000 sq. ft. (120×100 ft.). Larger exterior dimensions can be un-sprinklered if they have rated fire separations to reduce square foot dimensions under the 12,000 range.

If the exterior load-bearing walls are made totally of concrete and the mall building is only one story tall, tilt-up concrete construction can be suspected. These buildings are essentially held together by steel cable, which can fail at 800°F. Steel exterior load-bearing walls can have several sidings attached to the steel columns. Steel siding, brick, aluminum, or even wood can be applied. Concrete slabs similar to tilt-up construction can also serve as a siding.

Watch buildings go up in your community. Have the building officials notify your department of new construction starts and then drive by often. Know your buildings.

Roof assemblies

Roof assemblies for big box stores will generally be metal deck, membrane, or rubber. These will normally be over steel bar joist supports. Any air conditioning units on newer malls should have been planned and supported appropriately.

Roof assemblies for malls can be several types, depending on the age of the building. Older converted malls can have tar and gravel over plywood or plank roofing. Newer rubber or EPDM membrane roofs can also be applied to older plank or plywood roof sheathing. Be careful of added dead loads such as air conditioning units on older buildings. Try to ascertain if strengthening was added to the joist system to account for the added weight.

Newer lightweight malls will normally have metal deck, membrane, or rubber roofs. These will be installed over steel bar joist supports. Any air conditioning units on newer malls should have been planned and supported appropriately.

Floor assemblies

Floor assemblies for big box stores will almost always be concrete slab. Older converted malls will usually have floors made of wood or concrete. Coverings will be linoleum, tile, carpet, or painted concrete. Newer lightweight malls will have concrete floors as slabs or Q-decking supported by steel bar joists for multi-level malls. Tile or carpeting is usually the final floor covering.

Prioritizing Search in Large Malls and Big Box Stores

These occupancies are businesses and not intended for sleeping or use as a health care facility. They are not as prone to large loss of life fires as other occupancy types. However, due to their sheer size, if a fire starts during business hours, locating and removing civilians and employees becomes a high priority. Their size and the potential for large amounts of heat production from the materials contained inside means any consideration of conducting a search in zero visibility or high IDLH conditions must be taken very seriously. Advanced fire in these occupancies would indicate either catastrophic failure of the automatic sprinkler system or a fire of sufficient volume and intensity to overwhelm the system. Either event would put the structural stability of the building in extreme jeopardy.

Searching Large Malls and Big Box Stores

Neither big box stores nor enclosed malls pose a significant life hazard in the event of a fire due to the alert state of the occupants and early notification systems. Both types are expected to have built-in automatic fire protection. Even if the system is overwhelmed by fire, the audible alarm would most likely activate, alerting workers and shoppers.

Another factor weighs positively in the construction of these occupancies. They have very high ceilings, which will delay the banking down of smoke to the floor level. With good visibility and hose lines in place to hold or extinguish the fire, rapid searches can be conducted. I do not advocate conducting any search in an occupancy of this type with zero visibility. The likelihood of finding savable victims vs. the stability of the structure, potential amount of fire load, and the safety of fire crews does not weigh out in favor of search. The risk would simply not be worth the reward in attempting search in a big box or department store in zero visibility conditions.

Standard search

Due to the size and extremely complicated layout, I would not attempt a standard search in either of these occupancy types. You must constantly remain aware of your entry point. That must be established first in any attempt to conduct a search in these buildings. The simple definition of standard search removes this safety valve thus rendering it an unsafe option, in my opinion.

Team search

Team search is an option if a search is to be conducted in big box stores or malls. There are many columns located throughout these buildings, as well as storage and display racks of significant weight to use to wrap the search rope around in order to keep it taut and suspended off the floor.

As stated above, I would not attempt any search operation in zero visibility in one of these occupancies. However, I would never attempt a search in a big box store involved in fire *without* the aid of a search rope, in the event that conditions deteriorate rapidly. I realize I said *never.* With visibility, I would use a search rope and the team search method if it were determined that a search team should enter the building. In the event that visibility begins to worsen, the search rope is utilized to exit the building immediately upon noticing the increase of smoke. I would strongly consider abandoning any search if visibility in one of these occupancies becomes an issue.

Some additional thoughts on team search in big box stores and enclosed malls:

- Big box stores have aisles that can be used to maintain orientation along with the search rope.

- Anchor stores and enclosed malls have many columns you can use for tying the search rope. However, the display racks in most department stores are not of sufficient weight or mass to wrap. They are likely to simply fall over and have a tendency to be dragged across the floor.

- Only attempt team search with a search rope if you have trained—and trained often—using this procedure.

The oriented search

I would not consider conducting an oriented search in a big box store or enclosed mall without the aid of a 2½-in. hoseline to be used as a reference to the point of entry. Because of the size of the stores, a pre-connected 2½-inch hoseline will not get you very far inside the structure.

Maintaining continuity of search

I would not consider conducting a search unless I had visibility, so I would not consider maintaining continuity of search a huge problem. If I expected to visually locate several victims, I would establish a rescue group and instruct the same searcher to continue on with the search after locating a victim and bringing them to safety.

Again I must emphasize that I would not consider attempting to search in zero visibility in these occupancies. They are simply too big and complicated. Horizontal ventilation using many PPV fans or hovercraft, if available, would be an option to regain visibility in a reasonable amount of time.

Final thoughts

I'll say it again: It is not a good practice to commence a search in zero visibility in a big box store or similar occupancy. Depending upon the occupancy type and the likelihood of savable victims, this concept could also be applied in other occupancies such as strip malls, auto repair garages, furniture showrooms, and the similar large structures. It would be extremely easy for me to sit on any witness stand and justify this concept to a jury. I'm not sure I could sit on a witness stand and defend the opposite action. It's much harder to defend sending crews into these occupancy types and losing firefighters in the process, even if civilian lives are in danger. Seeing or hearing victims are one thing: endangering firefighter lives on hearsay is another.

Searching a big box store in zero visibility equates to one thing: an extremely large fire producing significant quantities of smoke to obscure visibility. If the entire structure is involved, it shows an obvious failure of the building's fire protection system. Any civilian that is neither seen nor heard is almost assuredly deceased, probably prior to our arrival. Don't let anyone else end up that way.

Final Thoughts

28

Ideally, after reading this text you have realized that there could be room for improvement in your procedures for fireground search. We can always improve on everything that we attempt. Please remember there isn't one perfect search for any one particular application. Variations may have to be ad-libbed mid-search.

I wish to conclude this text with a summary of thoughts on search and the tactical evolutions that support search.

A few very large departments can arrive on scene with sufficient personnel to accomplish multiple tasks concurrently. To the rest of the firefighting community, I offer the following list of evolutions to accomplish at every working fire:

- **Put the fire out.** If you arrive and have a working fire with the possibility of victims inside, but they are neither seen nor heard, your initial action should be to get a hoseline in a position that gets water between savable victims and the fire. This line should be pulled as close to the seat of the fire is possible and opened immediately to darken or extinguish the fire. The exception to this is when a parent or other family member tells you exactly where a victim is.

- **Vent**

  - ▷ **With the hose stream.** If no other crews are available and visibility is zero, attempt to vent with the hose stream using a fog pattern out the closest available window. Please pay particular attention to fire conditions and the location of the fire. This evolution may cause the fire to be drawn toward the nozzle and window area in a flare-up. If staffing permits, two or more firefighters can start a search. If conditions become too IDLH, the hoseline can be left where it is and the entire crew should exit the building to begin positive pressure ventilation. Once visibility improves, crews can enter the house, following the hoseline to the fire area. With improved visibility, they can then look for victims. If visibility is decent and the fire has been knocked down, the crew can leave a firefighter on the hoseline and the rest of the crew can begin search.

  - ▷ **Using PPV.** If another crew is available, positive pressure ventilation should be started as soon as the fire is located and knocked down.

- **Begin search efforts.** You should begin looking for victims as soon as visibility and conditions improve.

Again, I do not advocate commencing search in vacant buildings on the word of civilians and bystanders unless a victim is either seen or heard by fire personnel. That's not the same thing as arriving at a house fire where a parent tells you that there are kids are inside.

The face and body language of a civilian claiming to be a parent or family member will give you a strong indication of whether he or she is telling the truth. More often than not, when we arrive at a vacant building fire and a bystander tells us that they saw someone go into the building prior to the fire, the report proves false. Many firefighters have been killed and seriously injured attempting to rescue victims that were never there, especially in obviously vacant or unoccupied buildings.

* * *

I realize I have thrown a lot of different concepts and thoughts at you. I hope you can convert some of those thoughts and concepts into an improved fireground search, regardless of the occupancy type you face.

Index

P–Q

R

T